Laboratory Exercises in
ENVIRONMENTAL
GEOLOGY

Laboratory Exercises in
ENVIRONMENTAL
GEOLOGY

Second Edition

HARVEY BLATT
Hebrew University of Jerusalem

Boston, Massachusetts   Burr Ridge, Illinois   Dubuque, Iowa
Madison, Wisconsin   New York, New York   San Francisco, California   St. Louis, Missouri

McGraw-Hill

*A Division of The **McGraw·Hill** Companies*

LABORATORY EXERCISES IN ENVIRONMENTAL GEOLOGY

3 4 5 6 7 8 9 0 QPD/QPD   0 9 8 7 6 5 4 3 2 1

ISBN 0-697-28288-0

Publisher: *Edward E. Bartell*
Sponsoring Editor: *Lynne M. Meyers*
Developmental Editor: *Daryl Bruflodt*
Marketing Manager: *Lisa L. Gottschalk*
Project Manager: *Donna Nemmers*
Production Supervisor: *Laura Fuller*
Designer: *Mary L. Christianson*
Cover designer: *Barb Hodgson*
Photo research coordinator: *Carrie K. Burger*
Art Editor: *Renee A. Grevas*
Compositor: *York Graphic Services, Inc.*
Typeface: *10/12 Times Roman*
Printer: *Quebecor, Dubuque*

The credits section for this book is on page 177 and is considered an extension of the copyright page.

www.mhhe.com

CONTENTS

INTRODUCTION

The past decade has witnessed an explosion of interest in the relationship between geology and the everyday concerns of the average citizen. But why should someone who is not a professional geologist care about planet Earth? So what if the earth's surface is composed of plates, rather than an unbroken, homogeneous skin? Why should anyone care whether or not a rock layer contains oriented clay minerals? Will the fact that calcium carbonate is more soluble than quartz affect *your* life in any meaningful way?

The answers to these questions reveal that the average citizen can indeed be affected by such concerns. The fractured, platy character of the earth's crust, for example, affects the cost of house insurance in California and elsewhere. The presence of oriented clay minerals predisposes inclined rock layers to slide downhill in many parts of the United States. The difference in solubility between calcium carbonate and quartz causes some Florida homes to collapse into huge pits.

The relationship between geology and short-term human concerns (periods of no more than a few hundred years) is termed *environmental geology.* The purpose of this laboratory manual is to provide examples of how rocks and minerals exposed at the earth's surface and geological processes affect the natural environment.

At most schools, environmental geology is taught at the first-year level, commonly as an alternative to a standard Physical Geology course, and is designed for students who need to fulfill a college science requirement. The course also serves as an elective for students interested in the environment. Because of the dual function of this course, students in the class often have quite varied science backgrounds. Some have had a previous university science course, while others have had none, and might even have avoided science and mathematics in high school. To help accommodate the needs of both groups, the exercises in this laboratory manual typically contain more questions than can be answered within a standard two- to three-hour laboratory class, and the questions vary in difficulty. In addition, each exercise contains a mixture of hands-on and thought-provoking questions that deal with social, ethical, or political issues of environmental relevance. The instructor can use whichever questions seem most appropriate.

The manual also contains more than enough exercises for the normal 14 labs per semester, so instructors can choose the exercises most suitable for their geographic area.

For example, problems of floods and coastal erosion seem more immediate in Louisiana than in Idaho.

This manual is designed to be used in conjunction with textbooks on physical geology and environmental geology. For this reason I have avoided the lengthy theoretical discussions normally included in those texts. I have, however, described some of the basic principles that underlie each laboratory exercise.

These principles cannot be discussed without reference to size and distance. How far is a house from an unstable hillside? Is the hill slope steep or gentle? Scientists usually use a metric scale in answering such questions; engineers and other professionals use the familiar, nonmetric scale of feet and inches. Scale problems arise with temperature measurements. Scientists use the centigrade or Kelvin scale rather than the Fahrenheit scale more familiar to American students. To help students gain familiarity with various systems, this manual uses different scales in different exercises. Students also need to learn how to convert easily from one set of units to another; therefore, numerous conversion factors are included in Appendix A.

During the past 25 years, the increasing importance of environmental problems has led to the publication of many books that are important references for environmental concerns. Books that deal with geological influences on the environment include the following:

Blatt, H., 1997. *Our Geologic Environment.* New York, Prentice-Hall, 541 pp.

Coates, D. R., 1981. *Environmental Geology.* New York, John Wiley & Sons, 701 pp.

Coates, D. R., 1985. *Geology and Society.* New York, Chapman and Hall, 406 pp.

Costa, J. E., and Baker, V. R., 1981. *Surficial Geology.* New York, John Wiley & Sons, 498 pp.

Dennen, W. H., and Moore, B. R., 1986. *Geology and Engineering.* Dubuque, Iowa, Wm. C. Brown, 378 pp.

Garrels, R. M., Mackenzie, F. T., and Hunt, C., 1975. *Chemical Cycles and the Global Environment: Assessing Human Influences.* Los Altos, California, William Kaufmann, 206 pp.

Goudie, A., 1990. *The Human Impact on the Natural Environment,* 3rd ed. Cambridge, Massachusetts, MIT Press, 388 pp.

Griggs, G. B., and Gilchrist, J. A., 1983. *Geologic Hazards, Resources, and Environmental Planning,* 2nd ed. Belmont, California, Wadsworth, 502 pp.

Keller, E. A., 1996. *Environmental Geology,* 7th ed. New York, Prentice-Hall, 560 pp.

Legget, R. F., 1973. *Cities and Geology.* New York, McGraw-Hill, 624 pp.

Leveson, D., 1980. *Geology and the Urban Environment.* New York, Oxford University Press, 386 pp.

Lundgren, L., 1986. *Environmental Geology.* New York, Prentice-Hall, 576 pp.

McCall, G. J. H., DeMulder, E. F. J., and Marker, B. R. (eds.), 1996. *Urban Geoscience.* Rotterdam, Holland, A. A. Belkema, 273 pp.

Montgomery, C. W., 1998. *Environmental Geology,* 5th ed. Dubuque, Iowa, The McGraw-Hill Companies, Inc., 544 pp.

Murck, B. W., Skinner, B. J., and Porter, S. C., 1996. *Environmental Geology.* New York, John Wiley, 535 pp.

Pipkin, B. W., and Trent, D. D., 1997. *Geology and the Environment, 2d edition.* Belmont, California, Wadsworth Publishing Company, 522 pp.

Ward, K. (ed.), 1989. *Great Disasters.* Pleasantville, New York, Reader's Digest Association, 320 pp.

I hope students will finish this laboratory manual with a better understanding and appreciation of the ground beneath their feet. I encourage both faculty and student users of this manual to send suggestions for improving any of the exercises to either the author or the publisher.

INTRODUCTION TO THE SECOND EDITION

This second edition of *Laboratory Exercises in Environmental Geology* has been heavily revised based on suggestions made by users of the first edition. For example, the three exercises on rocks have been condensed into a single exercise to make room for additional exercises on environmental topics. Other exercises in the first edition were either eliminated or combined with others. New exercises in this edition include those on the nature of environmental geology, swelling soils, water pollution, mineral resources, air pollution, and alternative energy sources. The number of exercises has been cut from 24 to 21. In addition, reference lists have been updated and shortened and a glossary has been added.

Perhaps the most frequent complaint by users of the first edition was the level of mathematics. It was considerably greater than is traditional in geology laboratories. In response to this concern, most of the math has been either abbreviated or eliminated in this edition of the manual. Another objection to the first edition was the use in several exercises of chemical equations to illustrate what occurs in nature. Many of the students who take this course are not science majors and therefore do not have the necessary background to handle these equations. Hence, the amount of chemistry in this edition has been considerably reduced.

I again encourage the inclusion of field trips in the environmental geology course, either as a substitute for laboratory time or as weekend excursions. Governmental agencies and companies involved with environmental problems are usually more than happy to explain what they do and why. These include the U.S. Geological Survey and state geological surveys, groundwater companies, companies that operate sanitary landfills, mining companies, and companies that deal with sources of alternative energy, such as wind power and solar power.

As with the first edition, I hope students will finish their environmental geology course with an increased understanding of the world in which they live. If this laboratory manual contributes to this understanding, I will feel my writing effort was worthwhile. Once again, I encourage users of the manual to send their suggestions for improvement to the publisher.

R E V I E W E R S

I wish to extend a special thanks to the following people for their thoughtful comments and suggestions in prepublication reviews:

First Edition Reviewers:
John R. Huntsman
University of North Carolina at Wilmington

Rob Sternberg
Franklin and Marshall College

Mary P. Anderson
University of Wisconsin–Madison

Paul D. Nelson
St. Louis Community College at Meramec

George P. Merk
Michigan State University

James R. Lauffer
Bloomsburg University

William D. Nesse
University of Northern Colorado

Jim Constantopoulos
Eastern New Mexico University

David L. Ozsvath
University of Wisconsin–Stevens Point

Second Edition Reviewers:
Michael Whitsett
University of Iowa

Barbara Ruff
University of Georgia

Their contributions were enlightening, challenging, and encouraging throughout the development process.

E X E R C I S E 1

WHAT IS ENVIRONMENTAL GEOLOGY?

Geology is the study of the earth: its history, its rocks, its waters, its atmosphere, and the life that exists on it. These are our surroundings. This is our environment. So, in a sense, the expression *environmental geology* contains an unnecessary word: *environment.* Geology is by definition an environmental subject. This has not always been obvious, however. For most of the 20th century and until recently, most geologists were involved in the search for underground supplies of oil and gas. The connection to human concerns at the earth's surface, such as landslides, soil quality and food production, or the occurrence of floods, was not obvious. So the term *environmental geology* arose to indicate those parts of geology that were clearly related to those human concerns that deal with surface processes. Environmental geology is a fast-growing area of study, and employment opportunities are increasing explosively for environmental geologists.

Most of you in this laboratory don't intend to become professional geologists. But all people will benefit by knowing something about their surroundings. A large and growing proportion of America's domestic concerns are environmental issues that every citizen is asked to vote on. Sometimes the issues are local, such as whether it is necessary to develop a new sanitary landfill in which to place the town's garbage or to enlarge the city sewage plant. Will the location of the landfill pose a threat to the underground water supply that feeds the town's wells? Sometimes the environmental issues are national, such as the question of how much federal money (your tax dollars) should be spent to decrease air pollution or to control global warming. Does your congressional representative reflect your view, industry's view, or the government's

view? Other environmental issues whose outcome may affect you personally include the possibility of a flood in your area, acid rain on your newly painted house, the likelihood of a landslide on the neighboring hillside, and the stability of the soil your house stands on. The list is endless. All of these topics, and others, will be considered in this laboratory manual. They are also treated in your textbook in more detail.

INFORMATION

What do you need to know about the scientific aspects of our environment in order to have an informed opinion about environmental issues?

Earth Materials

Of what is the earth's surface composed (Figure 1.1)? What is a mineral? What is a rock? Do some minerals cause environmental problems? What is soil? How does soil form? Is all soil equally nourishing for crops? How does soil become polluted? Can soil pollution be prevented? How does well water get into the ground? Is it always safe to drink and, if not, why not?

Earth Locations

Which areas are environmentally at risk or dangerous? Is a landslide more likely here or there? Is a major flood likely where I live? Is sea level likely to rise and erode the beach and innundate my beach house (Figure 1.2)? Will there be an earthquake in my area soon? Surface mining of coal (strip mining) can cause water pollution. Where are the nearest coal deposits? The answers to these questions can be found only

Figure 1.1

Did the water in this flooded rice field originate in the mountains in the background? Why doesn't the water sink into the ground and disappear from view? What is the soil made of?

by referring to maps that show the shape of the land surface and maps that show the distribution of rocks and sediment.

Earth Underground

Most crops are irrigated, and most of the water for irrigation comes from underground. Will this underground water always be available? How do we know? Will the water flow faster if the rock layers below the ground are tilted toward my well? How serious a problem is pollution of my well water? Can polluted underground water be cleansed?

Figure 1.2

This beach house along the Atlantic shore in North Carolina is in danger of collapse during the next winter storm. Will the sandbags protect it? What else might be done? Should the house have been built in this location?

Why do earthquakes and volcanoes occur in some places but not in others? Is there a pattern to their locations that might be useful for predictions (Figure 1.3)? If so, what causes it? Is my house in danger?

Figure 1.3

Patterns of earthquake and volcanic activity based on historic records.

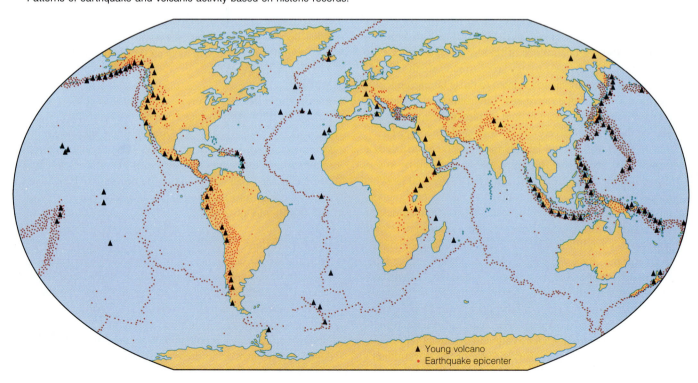

▲ Young volcano
• Earthquake epicenter

Figure 1.4

Aerial photograph showing the result of the most destructive landslide in the history of Hong Kong, June 18, 1972. Almost 26 inches of rain fell during the preceding two days, saturating and undermining the steep slope, which is composed of soil and underlying unlithified coarse sediment. The slope failure was 220 feet wide and destroyed a 4-story building and a 13-story apartment building; 67 people were killed. Hong Kong's population density and topographic relief make it impossible to completely avoid construction in unsafe areas.

Why do oil and natural gas occur where they do? Are we running out of these resources? Where and why? Are there any practical alternatives to these sources of energy that we rely on to fuel our industrial society?

The Air We Breathe

TV tells me there is some evidence that the climate is becoming warmer and wetter because of the *greenhouse effect*. What is the greenhouse effect, and can anything be done about it? And what is the *ozone hole?* For that matter, what is ozone? Will alternatives to petroleum help to keep the air clean? Why? Do I need to be concerned about radon gas entering my house from the soil below?

NUMBERS

Finally, there is the question of numbers. Numbers tell us

How far. Is the unstable slope near enough to threaten my house (Figure 1.4)? Am I located so close to the river that the next flood is likely to wipe me out?

How soon. Will the harmful chemical from the railroad accident sink into the ground and contaminate my underground water supply next week, next year, or in 50 years? When will the retreating shoreline become a problem for my coastal city?

How serious. Are the amounts of lead and arsenic in my drinking water too high for safety? Is the amount of soot in the air high enough to damage my lungs?

Relationships among things. Does the rise in air pollution parallel the increase in automobile use? Does the dumping of chemicals in the water upstream correlate with an observable increase in rectal cancer downstream? Is the relationship between these things strong enough to make us believe it is meaningful, or might it just be coincidental? Cigarettes and lung cancer? Dioxin and birth defects? Coal dust and emphysema?

Numbers are essential for evaluating the significance of environmental information, as will be evident in many of the exercises.

Figure 1.5

Spheres of the earth.

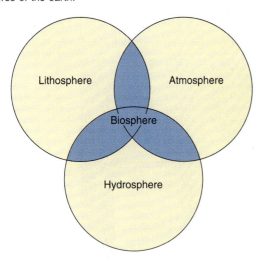

Figure 1.7

Shark in roof, Oxford U.K.

INTERRELATIONSHIPS

Everything is connected to everything else. Water (hydrosphere), rocks (lithosphere), air (atmosphere), and living matter (biosphere) are interrelated (Figure 1.5). Rain hits the land and dissolves rocks and minerals. This "impure" water is taken in by plant roots, which use many of the impurities as nutrients. However, some impurities are pollutants to both the plant and to the animals that eat the plant. And, of course, humans drink the stream water that has flowed through the surface rocks and soil.

Plant roots are rooted in soil ("dirt") that was produced by the rotting of rocks caused by chemical reactions between rock (lithosphere), rain (hydrosphere), and atmospheric gases. The soil also contains organic material (biosphere) from dead and decomposing plants and animals (plant leaves, worm skins, and other carcass parts you'd rather not think about).

Figure 1.6

What should be done about the colored plumes coming from these smokestacks? Imagine thousands of factories worldwide spewing these fumes. Do you want to breathe this stuff? Will your lungs suffer?

Some of the water that fell as rain or snow evaporates directly back to the atmosphere to raise the humidity. Some runs off on the ground surface toward the ocean. Also entering the air we breathe are various harmful chemicals that pour from industrial smokestacks (Figure 1.6). These include the gases that are implicated in global warming.

What we do to one thing we do to everything.

METHODOLOGY

The goal of science is to discover the interrelationships among the objects in our physical world. Environmental geologists, like other scientists, approach problems in a certain way, commonly called the *scientific method*.

1. A scientist observes a phenomenon. This may be water flowing in a stream, a building shaking in an earthquake, a mass of rock hurtling downslope toward a village below, or some unexpected event (Figure 1.7).
2. The scientist establishes a *hypothesis* to explain the observation. A hypothesis is a statement of the scientist's first guess at the explanation for the observed event.
3. The scientist makes additional observations or measurements or conducts experiments to test the hypothesis. Every effort is made to exclude personal bias from the observations.

4. The scientist analyzes the results of all observations and experiments to determine whether they are expected or are consistent with the hypothesis. After a sufficiently large number of observations have been made and found to be consistent with the hypothesis, the scientist becomes confident enough in the hypothesis to upgrade the explanation to the status of a *theory*. After much additional testing over an extended period of time, usually many years, the theory may be upgraded to the status of a *law*.

5. Based on observations, the scientist looks for patterns. These patterns are used to create models that may explain how parts of nature work. The models of science may be mathematical, such as the law of gravity:

$$F = c\,\frac{m_1 m_2}{r^2}$$

where F = force of gravity
   c = a universal constant
   m_1 = mass (density × volume) of the first object
   m_2 = mass of the second object
   r = distance between the centers of the two masses

The models may also be physical, such as the internal structure of a mineral represented by a few colored balls and wires (Figure 1.8). Earth science, or environmental science, is basically a detective story with the scientist as the detective and the earth as the mystery.

Problems

1. Look in the telephone book to see whether any geologists are listed. If so, is there a subheading for environmental geologists? Are there any companies that deal with environmental issues, such as water testing and remediation, garbage disposal, earthquake-proofing of houses, or the manufacture of wind turbines or solar panels?

2. Does your state have a certification program for geologists, or can anyone claim expertise? Ask a geologist or the office of your state's geological survey. (Check the phone book again.)

3. What happens to the garbage you put in your trash can or dumpster that is picked up by the city garbage truck? Who handles this chore for your town? Where does the truck take it? To a big hole in the ground outside of town (a landfill) or to an incinerator to be burned? Is the disposal area filling up? How much garbage is collected each day, and how large is the hole it is put in? What does the town plan to do when the hole is full?

4. Who is responsible for monitoring the amount and purity of the drinking water in your town? How often is the water checked for contaminants? What contaminants does the person responsible look for?

5. Suppose you were worried about a geologic hazard in your town, such as a flood in the river that runs through the city, a volcano 10 miles away that is emitting plumes of smoke, flaking asbestos in city buildings, or an unpleasant taste and color in your drinking water. Where would you go for information and advice about such matters?

6. World population has reached 6 billion (6,000,000,000) people, with 100 million more (100,000,000) added yearly. Most of them are in the poorer areas of the world. Perhaps more frightening, the percentage of people living in urban areas is growing and has now reached about 50%. What types of environmental problems might these trends cause or make worse?

7. Consider a common object such as a pencil. What resources or materials from the earth were required to manufacture it? Which of them do you think might be renewable or inexhaustible, and which might we eventually run out of?

8. Suppose you wanted to conduct a survey to determine the views of those in your community about environmental issues. Prepare a list of five questions you would ask. Explain how you would choose the people to survey to get a balanced and fair representation of the "average person." What factors would you need to consider?

Further Reading/References

Blatt, Harvey, 1997. *Our Geologic Environment.* New York, Prentice-Hall, 541 pp.

Montgomery, Carla W., 1998. *Environmental Geology,* 5th ed. Dubuque, Iowa, WCB/McGraw-Hill, 544 pp.

Keller, Edward A., 1996. *Environmental Geology,* 7th ed. New York, Prentice-Hall, 560 pp.

Pipkin, Bernard W. and D. D. Trent, 1997. *Geology and the Environment.* Belmont, California, Wadsworth Publishing Company, 544 pp.

Figure 1.8

Atomic structure of halite, ordinary table salt (sodium chloride, NaCl). The large spheres are chlorine atoms, the smaller spheres are sodium atoms.

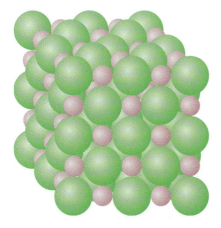

2 EXERCISE

MINERALS

Although our planet is composed of 90 chemical elements, they are not present in equal amounts. Eight elements form 99.4% by weight of the *crust*—the upper 30 miles under the land surface—while the other 82 elements total only 0.6%. The abundant elements are

Element	Weight %
Oxygen	46.4
Silicon	28.2
Aluminum	8.3
Iron	5.6
Calcium	4.1
Sodium	2.4
Magnesium	2.3
Potassium	2.1
All others	0.6

The composition and relative abundance of minerals reflect the chemical composition of the crust. The most abundant minerals are composed largely of oxygen, silicon, and aluminum.

Also noteworthy is the fact that most of the economically important elements are not among the eight most abundant elements. For example, titanium makes up only 0.57% of the crust of the earth, manganese is only 0.09%, chromium is only 0.01%, and other elements such as nickel, copper, and lead are even less abundant. Concentrations of most of these elements are uncommon—and are becoming even less so as our industrial civilization expands. Substitutes for most of them have yet to be found or synthesized.

Elements tend to combine into larger groupings because of their electronic structures. The change in electron distribution that results when elements combine determines the physical and chemical characteristics of the materials (e.g., minerals) produced. Unfortunately, however, the new chemical properties cannot be predicted from the properties of the individual, uncombined elements. For example, at room temperature sodium (Na) is a metal and chlorine (Cl) is a gas. But when the two combine as sodium chloride (NaCl, halite), they produce a solid—ordinary table salt. The properties of table salt (including very high solubility in water or in steak juice) are determined by the distribution of electrons in this sodium chloride aggregate, just as the properties of the uncombined sodium (metallic appearance, high melting temperature) and chlorine (irritating odor, greenish yellow color) are determined by *their* electronic structures.

Just as the properties of solid NaCl differ from those of uncombined sodium and chlorine, so do they also differ from the properties of sodium and chloride ions (charged atoms) dissolved in water. The "salty" taste sodium ions create in water (or saliva) is well known. The arrangement of electrons around atomic nuclei underlies the physical and chemical properties of all materials.

Two properties of great importance to environmental scientists studying minerals are hardness and solubility. Quartz (SiO_2) is very hard and relatively insoluble in water; calcite ($CaCO_3$) is soft and moderately soluble in water;

halite (NaCl) is very soft and very soluble in water. The importance of these chemical properties to drinking-water quality and to building construction in humid climates is obvious.

IMPORTANT PROPERTIES OF MINERALS

A mineral is a naturally occurring, inorganic solid with a regular, periodic internal structure and a fairly definite chemical composition (Table 2.1). Because of this fixed internal structure and chemical composition, the physical and chemical properties of a mineral are constant and can be used to identify it. The following properties are those most useful in mineral identification:

1. *Hardness,* defined as the ability of a mineral to resist abrasion, is determined by scratching the mineral with an object of known hardness. Harder minerals scratch softer ones. Geologists use a hardness scale devised in 1824 by a German mineralogist, Friedrich Mohs, and known as the *Mohs hardness scale* (Table 2.2). Surface alteration can decrease hardness; therefore, hardness must be determined on a fresh mineral surface.

2. *Cleavage.* The strength of chemical bonds in a mineral differs for different pairs of elements. Because of this difference, and because the elements occur in fixed positions in its crystal structure, a mineral can have some planar surfaces across which bonding is weaker. When hit, the mineral tends to break along these weaker planes, which are called *cleavage surfaces* (Figure 2.1). Minerals can have one, two, three, four, or six different cleavage directions, and these can be diagnostic for the mineral. For example, micas have one cleavage direction and cleavages occur as sheets. In a highly micaceous rock, slippage and slope failure tend to parallel oriented groups of mica flakes.

Some minerals do not show cleavage, either because their cleavage surfaces are poorly developed or because their chemical bonds are nearly equal in all directions. Quartz is a mineral with no obvious cleavage.

Cleavage faces should not be confused with crystal faces. Cleavage faces are planar surfaces of preferential breakage that reflect planes of weakness in a crystal structure. Crystal faces, in contrast, are external planar surfaces that form as a mineral grows from a solution; they reflect the geometry of the internal structure. Only rarely is the shape of cleavage fragments of a mineral the same as the shape of its fully formed crystals.

TABLE 2.1

Common Rock-Forming Minerals

Abundant

Mineral	Chemical composition
Quartz	SiO_2
Orthoclase feldspar	$KAlSi_3O_8$
Plagioclase feldspar	$NaAlSi_3O_8$ to $CaAl_2Si_2O_8$
Biotite mica	hydrous K, Fe, Mg, Al silicate
Muscovite mica	$KAl_3AlSi_3O_{10}(OH)_2$
Hornblende	Ca, Na, Mg, Fe, Al silicate
Augite	Ca, Mg, Fe, Al silicate
Olivine	$(Mg, Fe)_2SiO_4$
Chlorite	hydrous Mg, Fe, Al silicate
Illite clay	hydrous K, Al, Fe, Mg silicate
Montmorillonite clay	hydrous Na, Ca, Al, Fe, Mg silicate
Kaolinite clay	$Al_2Si_2O_5(OH)_4$
Calcite	$CaCO_3$
Dolomite	$CaMg(CO_3)_2$
Gypsum	$CaSO_4 \cdot 2H_2O$
Halite	NaCl
Hematite	Fe_2O_3

Less Abundant but Still Common

Mineral	Chemical composition
Garnet	Fe, Mg, Ca, Al silicate
Kyanite	Al_2SiO_5
Sillimanite	Al_2SiO_5
Staurolite	Fe, Mg, Al silicate
Epidote	Ca, Fe, Al silicate
Magnetite	Fe_3O_4
Illmenite	$FeTiO_3$
Pyrite	FeS_2
Graphite	C

TABLE 2.2

Mohs Hardness Scale

Relative Hardness	Index Mineral	Common Objects
10	Diamond	
9	Corundum	
8	Topaz	
7	Quartz	Steel file—6.5
6	Orthoclase	
5	Apatite	Glass, knife, nail—5.5
4	Fluorite	
3	Calcite	Copper penny—3.0
		Fingernail—2.5
2	Gypsum	
1	Talc	

Figure 2.1

Cleavage patterns of minerals. Few common minerals have more than three cleavage directions.

Number of cleavage directions	Shape	Sketch
0 No cleavage, only fracture	Irregular masses (quartz)	
1	Flat sheets (micas)	
2 at 90°	Elongated form with rectangular cross-section (prism: spodumene)	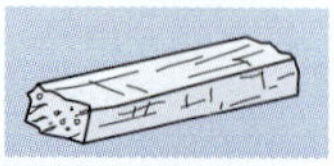
2 not at 90°	Elongated form with parallelogram cross-section (prism: hornblende)	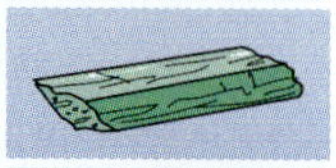
3 at 90°	Cube (halite)	
3 not at 90°	Rhombohedron (calcite)	
4	Octahedron (fluorite)	
6	Dodecahedron (sphalerite)	

Crystals normally are bounded by many more surfaces intersecting at different angles than are cleavage fragments. Among the few examples of cleavage fragments identical to fully formed crystals are halite cubes, galena cubes, and dolomite rhombohedra.

3. *Color.* Many of the abundant or common rock-forming minerals have distinctive colors that are useful for identification. Some minerals (particularly quartz, fluorite, and calcite) can occur in a wide variety of colors, but one color is the most common. Colors generally result from the presence of impurities that selectively absorb wavelengths of light entering the mineral. Those wavelengths that are not absorbed give the mineral its color.

4. *Streak,* the color of the mineral powder, is determined by powdering a sample, usually by scratching it across a piece of unglazed porcelain (hardness 7). The mineral must be softer than the porcelain, or it will scratch (powder) the porcelain rather than being powdered itself. Most nonmetallic minerals have a white or colorless streak; hence, streak is not a helpful diagnostic tool for the abundant minerals, almost all of which are nonmetallic. Streak is more helpful for identifying metallic minerals, many of which are of great economic importance.

5. *Luster* is the appearance of a fresh mineral surface in reflected light. A mineral that appears metallic is said to have a *metallic* luster. Nonmetallic mineral surfaces can be *vitreous* (glassy luster), *resinous, pearly, silky, dull,* or *earthy.* As examples, quartz is vitreous, sphalerite is resinous, talc is pearly, satin spar gypsum is silky, and microcrystalline hematite is dull or earthy.

6. *Specific gravity* is the ratio between the weight of a mineral and the weight of an equal volume of water. Most minerals have specific gravities of between 2.6 and 3.5. In general, the higher the content of heavy elements such as iron and lead, the higher the specific gravity of the mineral. For example, the specific gravity of magnetite (Fe_3O_4) is 5.2; that of galena (PbS) is 7.5. Gold, at 19, has the highest specific gravity of any mineral.

7. Other physical properties are sometimes useful in mineral identification (Table 2.3). For example, calcite is the only important mineral that dissolves in cold, dilute hydrochloric acid. Magnetite is the only mineral attracted to a small hand magnet, while halite has a unique salty taste. Micas are elastic when bent. Minerals usually occur as aggregates composed of many crystals, rather than as single crystals. The most diagnostic properties of the common minerals are listed in Table 2.4.

ECONOMIC MINERALOGY

A group of about 25 minerals includes probably 99%, by volume, of those present in the earth's crust. These abundant minerals predominate in the common rocks. But more than 3,500 different minerals are known, and many of the less common ones are important sources of chemical elements needed in our industrial civilization (Table 2.5). Others are used for decorative purposes, such as gems in jewelry. Many cities have gem and mineral societies that organize meetings and exhibitions to display and trade semi-precious gemstones.

ENVIRONMENTAL ASPECTS OF MINERALS

Some minerals can cause environmental problems because of their physical properties and chemical compositions—for example, calcite, halite and gypsum, pyrite, and clay minerals.

Calcite is the essential mineral in *limestone* and is also one of the more easily dissolved of the abundant minerals. Because limestones are such widely distributed rocks, dissolution of the ground surface and shallow subsurface is a common phenomenon in many areas. Rainwater seeps into cracks in the limestone and within a few hundred to a few thousand years can create large holes in the rock. When this occurs at shallow depths, perhaps a few tens of feet below the ground surface, it produces caverns such as Carlsbad Cavern in New Mexico. The ground above such a cavern can collapse into it, carrying with it buildings, automobiles, and even people. Numerous cases of ground collapse have

TABLE 2.3

Mineral Classification Chart

Metallic Luster		Other Characteristics	Mineral
	Gray streak	Perfect cubic cleavage; H = 2.5; heavy, sp. gr. = 7.6; silver gray color	Galena PbS
	Black streak	Magnetic; black to dark gray; H = 6; sp. gr. = 5.2; commonly occurs in granular masses; single crystals are octahedral	Magnetite Fe_3O_4
	Gray to black streak	Steel gray; soft, smudges fingers and marks paper, greasy feel; H = 1; sp gr. = 2; luster may be dull	Graphite C
	Greenish black streak	Golden yellow color; may tarnish purple; H = 4; sp gr. = 4.3	Chalopyrite $CuFeS_2$
		Brass yellow; cubic crystals; common in granular aggregates; H = 6–6.5; sp. gr. = 5; uneven fracture	Pyrite FeS_2
	Reddish brown streak	Steel gray, black to dark brown, red to red-brown streak; granular, fibrous, or micaceous; single crystals are thick plates; H = 5–6; sp. gr. = 5; uneven fracture	Hematite Fe_2O_3
	Yellow-brown streak	Yellow, brown, or black; hard structureless or radial fibrous masses; H = 5–5.5; sp. gr. = 3.5–4	Limonite $FeOOH \cdot nH_2O$

Nonmetallic Luster—Dark Color

Harder than glass		Other Characteristics	Mineral
	Cleavage prominent	Cleavage—2 directions nearly at 90°; dark green to black; short prismatic 8-sided crystals; H = 6; sp. gr. = 3.5	Pyroxene Group Complex Ca, Mg, Fe, Al silicates
		Cleavage—2 directions at approximately 60° and 120°; dark green to black or brown; long, prismatic 6-sided crystals; H = 6; sp. gr. = 3.35	Amphibole Group Complex Na, Ca, Mg, Fe, Al silicates
		White to gray; good cleavage in two directions at approximately 90°; striations on cleavage planes; H = 6; sp. gr. = 2.62–2.76	Plagioclase Feldspar $NaAlSi_3O_8$ to $CaAl_2Si_2O_8$
	Cleavage absent	Various shades of green; sometimes yellowish; commonly occurs in aggregates of small glassy grains; transparent to translucent; glassy luster; H = 6.5–7; sp. gr. = 3.5–4.5	Olivine $(Mg, Fe)_2SiO_4$
		Red, brown, or yellow; glassy luster; conchoidal fracture resembles poor cleavage; commonly occurs in well-formed 12-sided crystals; H = 7–7.5; sp. gr. = 3.5–4.5	Garnet Group Fe, Mg, Ca, Al silicates
		Conchoidal fracture; H = 7; gray to gray-black; vitreous luster; sp. gr. = 2.65	Quartz SiO_2

TABLE 2.3—*Continued*

Softer than glass	**Cleavage prominent**	Brown to black; 1 perfect cleavage; thin, flexible, and elastic when in thin sheets; H = 2.5–3; sp. gr. = 3–3.5	Biotite Hydrous K, Fe, Mg, Al silicate
		Green to very dark green; 1 cleavage direction; commonly occurs in foliated or scaly masses; nonelastic plates; H = 2–2.5.; sp. gr. = 2.5–3.5	Chlorite Hydrous Mg, Fe, Al, silicate
		Yellowish brown; resinous luster; cleavage in 6 directions; yellowish brown or nearly white streak; H = 3.5–4; sp. gr. = 4	Sphalerite ZnS
		Four perfect cleavage directions; green through deep purple; transparent to translucent; cubic crystals; H = 4; sp. gr. = 3	Fluorite CaF$_2$
	Cleavage absent	Red, earthy appearance; red streak; H = 1.5; sp. gr. = 5.26	Hematite Fe$_2$O$_3$ (earthy variety)
		Yellowish brown streak; yellowish brown to dark brown; commonly in compacted earthy masses; H = 1.5; sp. gr. = 3.6–4.0	Limonite FeOOH · nH$_2$O

Nonmetallic Luster—Light Color

Harder than glass	**Cleavage prominent**	Good cleavage in 2 directions at approximately 90°; commonly flesh-colored to dark pink; pearly to vitreous luster; H = 6; sp. gr. = 2.6	Orthoclase feldspar: KAlSi$_3$O$_8$
		Good cleavage in 2 directions at approximately 90°; white to gray; striations on some cleavage planes H = 6; sp. gr. = 2.62–2.76	Plagioclase feldspars: NaAlSi$_3$O$_8$ to CaAl$_2$Si$_2$O$_8$

	Cleavage absent	Conchoidal fracture; transparent to translucent; vitreous luster; 6-sided prismatic crystals terminated by 6-sided triangular faces in well-developed crystals; vitreous to waxy; colors range from milky white, rose pink, violet, to smoky gray; H = 7; sp. gr. = 2.65	Quartz
			SiO$_2$ (silica)
			Varieties:
			  Milky; Smoky;
			  Rose; Amethyst

		Conchoidal fracture; variable color; translucent to opaque; dull or clouded luster; colors range widely from white, gray, red, to black; H = 7; sp. gr. = 2.65	Microcrystalline
			Quartz SiO$_2$
			Varieties:
			  Agate; Flint; Chert;
			  Jasper; Opal (amorphous)

		Perfect cubic cleavage; salty taste; colorless to white; soluble in water; H = 2–2.5; sp. gr. = 2	Halite NaCl
		Perfect cleavage in 1 direction; poor in 2 others; white; transparent; nonelastic; H = 2; sp. gr. = 2.3 Varieties: Selenite: colorless, transparent Alabaster: aggregates of small crystals Satin spar: fibrous, silky luster	Gypsum CaSO$_4$ · 2H$_2$O

TABLE 2.3—*Continued*

Softer than glass — *Cleavage prominent*	Perfect cleavage in 3 directions at approximately 75°; effervesces in HCl; colorless, white, or pale yellow, rarely gray or blue; transparent to opaque; H = 3; sp. gr. = 2.7	Calcite $CaCO_3$ (fine-grained crystalline aggregates form limestone and marble)
	Three directions of cleavage as in calcite; effervesces in HCl only if powdered; color variable but commonly white or pink; rhomb-faced crystals; H = 3.5–4; sp. gr. = 2.8	Dolomite $CaMg(CO_3)_2$
	Good cleavage in 4 directions; colorless, yellow, blue, green, or violet; transparent to translucent; cubic crystals; H = 4; sp. gr. = 3	Fluorite CaF_2
	Perfect cleavage in 1 direction, producing thin, elastic sheets; transparent and colorless in thin sheets; H = 2–3; sp. gr. = 2.8	Muscovite $KAl_3AlSi_3O_{10}(OH)_2$
	Green to white; soapy feel; pearly luster; foliated or compact masses; one direction of cleavage forms thin scales and shreds; H = 1; sp. gr. = 2.8	Talc $Mg_3Si_4O_{10}(OH)_2$
Cleavage absent	White to red; earthy masses; crystals so small no cleavage visible; becomes plastic when moistened; earthy odor; soft; H = 1.2; sp. gr. = 2.6	Kaolinite $Al_2Si_2O_5(OH)_4$

occurred in the midcontinent and in populated areas of Florida and other southeastern states that have widespread, abundant limestone.

Halite and gypsum are much more soluble than calcite, but also much less abundant, so they cause fewer cases of ground collapse. However, they pose a great danger to the water supply in many regions, because even small amounts of sodium or sulfate ions in water can produce a salty taste (sodium) or a laxative effect (sulfate). These problems are so severe in some areas, such as parts of west Texas, that the residents must purchase bottled water imported from other regions.

Pyrite causes environmental problems in some of the mining areas where it is particularly abundant. The mineral dissolves and combines with oxygen in the atmosphere to form hydrogen ions, a major cause of acidic stream water.

$$4\ FeS_2 + 15\ O_2 + 8\ H_2O \rightarrow 2\ Fe_2O_3 + 8\ SO_4^{-2} + 16\ H^+$$

pyrite oxygen water hematite sulfate hydrogen
                                           ion   ion

Such waters, called *acid mine drainage,* cause serious pollution problems.

Clay minerals (kaolinite, montmorillonite, and illite) and, to a lesser extent, micas can trigger environmental problems because their crystal structure causes easy cleavage. The sheetlike structure of these minerals permits rainwater to penetrate between the sheets, causing the sheets to swell and slip. Houses and other structures built on clay-rich rocks such as *shales* can have their foundations cracked by swelling clays. Buildings constructed on slopes underlain by shales sometimes slide downhill when the shale becomes saturated with water, a common phenomenon in hilly areas such as those of southern California.

Problems

1. Identify each of the minerals provided, specifying the particular physical properties you used in the identification. Be specific (e.g., green color rather than simply "color," two cleavages at right angles rather than only "cleavage").

2. The chemical formulas of solid compounds such as minerals are expressed by the smallest number of each type of atom necessary for there to be no net charge. By convention, the positively charged ion is listed first. For example, sodium chloride, the mineral halite, is written in chemical symbols as NaCl. Write the chemical formulas of each of the following minerals:
   a. sylvite, potassium chloride
   b. quartz, silicon dioxide
   c. corundum, aluminum oxide
   d. hematite, iron oxide (iron is +3)

3. Biotite mica and muscovite mica have identical crystal structures. Why, then, are they considered different minerals? Is this reason reflected in the criterion you used to distinguish between them? Why or why not?

4. The musical stage play *Oklahoma* contains a mention of a surrey with "isinglass windows you can roll right down." Isinglass is an obsolete name for one of the minerals in your tray. Identify it and state the physical

TABLE 2.4

Diagnostic Characteristics of the Common Minerals

Mineral	Characteristics
Quartz	Transluscent; conchoidal fracture; hardness 7
Orthoclase feldspar	Right-angle cleavage; commonly pink; hardness 6
Plagioclase feldspar	Right-angle cleavage; commonly white, striated
Biotite mica	Sheet structure and cleavage; dark green-black; hardness 2.5
Muscovite mica	Sheet structure and cleavage; transluscent; hardness 2–2.5
Hornblende	Two cleavages at 56° and 124°; elongate; dark green to black
Augite	Two cleavages at 87° and 93°; black
Olivine	Light green; conchoidal fracture
Chlorite	Sheet structure and cleavage; green; hardness 2–2.5
Clay minerals	Microscopic crystals occurring as aggregates; hardness 2; kaolinite is white, montmorillonite and illite are green
Calcite	Dissolves with effervescence in dilute HCl; rhombohedral cleavage; normally white but other colors possible
Dolomite	Same as calcite but does not dissolve/effervesce unless powdered
Gypsum	Hardness 2; usually transluscent to white but other colors possible; three unequally good cleavages
Halite	Salty taste; three right-angle cleavages; hardness 2.5
Hematite	Red-brown streak; metallic luster in visible crystals; earthy when microcrystalline
Garnet	Usually red but other colors possible; hardness 7; commonly has 12-sided crystal outlines
Kyanite	Bladed crystals; bluish; hardness 5 parallel to crystal length, 7 normal to length; vitreous-pearly luster
Sillimanite	Long slender crystals often in parallel groups; frequently fibrous; hardness 6–7
Staurolite	Red-brown to black; resinous to vitreous luster; hardness 7–7.5; common interpenetrating right-angle twins
Epidote	Pistachio green; one perfect cleavage; hardness 6–7
Magnetite	Highly magnetic; black; hardness 6
Ilmenite	Like magnetite but nonmagnetic
Pyrite	Brassy yellow; typically in cubic crystals; streak greenish or brownish black
Graphite	Sheet structure and cleavage; readily soils fingers and marks paper (pencil "lead"); greasy feel; black

property that enabled it to be used in lieu of glass, which was very expensive in the 1800s.

5. How might you distinguish between the smooth surfaces of a perfectly formed crystal and the cleavage surfaces of the same mineral?

6. Graphite is commonly used as a lubricant. Explain why this is possible, in terms of a visible physical property of the mineral.

7. What product in your house or dormitory might be made from each of the following minerals?
   a. graphite
   b. calcite
   c. garnet
   d. halite
   e. gypsum
   f. quartz

8. Most caves are formed in limestone, a rock composed entirely of the mineral calcite.
   a. Give two reasons why this is true.
   b. Quartz and feldspar are the two most abundant minerals exposed at and near Earth's surface. Yet there are no commercial caves formed in rocks composed of these minerals. Why not?

TABLE 2.5

Ore and Gem Minerals

Element Recovered	Major Ore Mineral	Chemical Formula
Antimony	Stibnite	Sb_2S_3
Arsenic	Orpiment, Realgar	As_2S_3, AsS
Beryllium	Beryl	$Be_3Al_2Si_6O_{18}$
Chromium	Chromite	$FeCr_2O_4$
Cobalt	Cobaltite	$(Co, Fe)AsS$
Copper	Chalcopyrite	$CuFeS_2$
Iron	Hematite	Fe_2O_3
Lead	Galena	PbS
Manganese	Pyrolusite	MnO_2
Mercury	Cinnabar	HgS
Molybdenum	Molybdenite	MoS_2
Nickel	Pentlandite	$(Fe, Ni)_9S_8$
Tin	Cassiterite	SnO_2
Titanium	Rutile	TiO_2
Tungsten	Wolframite, Scheelite	$(Fe, Mn)WO_4$, $CaWO_4$
Uranium	Carnotite	$K_2(UO_2)_2(VO_4)_2 \cdot 3H_2O$
Vanadium	Carnotite	$K_2(UO_2)_2(VO_4)_2 \cdot 3H_2O$
Zinc	Sphalerite	ZnS
Zirconium	Zircon	$ZrSiO_4$
Native (uncombined) elements		Gold, platinum, silver, sulfur

Major Gem Minerals

Aquamarine (blue beryl)		$Be_3Al_2Si_6O_{18}$
Chrysoberyl		$BeAl_2O_4$
Diamond		C
Emerald (green beryl)		$Be_3Al_2Si_6O_{18}$
Garnet (common)		$Fe_3Al_2(SiO_4)_3$
Jade		$NaAlSi_2O_6$
Lapis lazuli		$(Na, Ca)_4(AlSiO_4)_3(SO_4, S, Cl)$
Olivine		$(Mg, Fe)_2SiO_4$
Opal		$SiO_2 \cdot H_2O$
Ruby (red corundum)		Al_2O_3
Sapphire (blue corundum)		Al_2O_3
Topaz		$Al_2SiO_4F_2$
Tourmaline		Borosilicate of variable composition
Turquoise		$CuAl_6(PO_4)_4(OH)_8 \cdot 4H_2O$
Zircon		$ZrSiO_4$

Plagioclase showing typical white color and striations.

Hornblende crystals. Hornblende usually occurs as sand-size crystals in rocks and is identified by its cleavage.

Orthoclase (microcline) showing typical salmon color.

Muscovite mica showing excellent sheetlike cleavage and translucency. Biotite mica is dark green.

Quartz crystal with striations.

Augite, which usually occurs as sand-size crystals, is identified by its cleavage.

Garnet crystals and white plagioclase feldspar.

Gypsum (hardness of 2) is easily scratched by a fingernail (hardness of 2 1/2).

Calcite. Note double refraction and rhombohedral cleavage.

Cubic and rectangular cleavage fragments of halite.

Pyrite showing brassy yellow color.

Galena. Showing metallic luster and cubic cleavage.

Fluorite. Two octahedral crystals and one cubic cleavage fragment.

Worksheet for Minerals

Sample	Hardness	Cleavage (number and angular relationship)	Color	Streak	Luster	Other Distinctive Properties	Name of Mineral

3 EXERCISE

ROCKS

Environmental geologists are interested in rocks for six reasons.

1. Some rocks decompose to yield nutrient-rich soils for plants, while others yield relatively unproductive soils that must be heavily fertilized (Exercise 7).

2. Rocks may contain minerals that swell when wet, making building foundations unstable (Exercise 8).

3. Some rocks fall apart or dissolve easily, leading to landslides, collapse of buildings, and erosion of coastlines (Exercises 9, 15).

4. Rocks may contain interconnected holes through which may flow water, oil, or solutions that drip through garbage disposal areas (Exercises 12, 13, 14, 19).

5. Some rocks are very susceptible to chaotic behavior during earthquakes, leading to almost certain destruction of structures built on them (Exercise 16).

6. Rocks may contain economically valuable minerals (Exercises 18, 20).

CLASSIFICATION OF ROCKS

Geologists group rocks into three categories based on the processes that form them. These three categories are independent of environmental concerns. A rock in any of the three may have environmental significance.

Igneous rocks are formed by solidification from molten material. One well-known example is lava that pours from a volcano and solidifies as it cools while flowing down the flanks of a volcano. Much molten rock material solidifies below the ground, although much more slowly.

Sedimentary rocks are formed at Earth's surface. Fragmental sediment such as sand becomes buried as new sediment is deposited on top of it. While it is buried, water flows through it and deposits minerals that cement the sand grains into a rock. Sedimentary rock may also precipitate entirely from water. Rock salt (halite) is an example.

Metamorphic rocks are formed from either of the other two types or from another metamorphic rock by heat and pressure. Temperatures of metamorphism range from 300–700° C.

The conversion of any one of the three types of rocks into one of the others has been occurring continuously since the earth was formed about 4.5 billion years ago. It is also happening right now, starting at a depth of perhaps 10 miles under your feet. The transitions are illustrated in Figure 3.1. It is called the *rock cycle*.

HOW TO TELL ONE TYPE FROM ANOTHER IN HAND SPECIMEN

There are no infallible criteria that separate small pieces (hand specimens) of the three groups of rocks because the groups grade into one another in appearance. There are exceptions to all of the criteria given below, but the criteria are usually reliable. Some criteria are textural, others mineralogic.

1. Does the rock contain fossils, evidence of living organisms? Perhaps there is a piece of clam shell or imprint of a leaf. Only sedimentary rocks

Figure 3.1

The rock cycle.

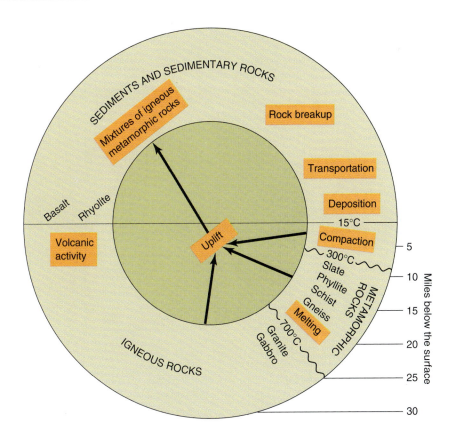

contain fossils because living organisms cannot survive at temperatures of more than 300° C.

2. Is the rock fragmental, composed of rounded grains held together by a cement of some sort? Only sedimentary rocks are fragmental.

3. Is there a wide variation in grain size of a single mineral, such as quartz? If so, the rock is sedimentary.

4. Is the rock composed entirely of calcite or dolomite? If so, it is either sedimentary or metamorphic. In a metamorphic rock composed of carbonate minerals, the crystals are normally much larger than in a sedimentary rock. In an unfossiliferous limestone, the crystals are commonly either too small to see or are barely visible.

5. Does the rock contain more than 40–50% feldspar? If so, it is either igneous or metamorphic.

6. Does the rock contain noticeable amounts of visibly micaceous minerals in parallel orientation? If so, the rock is metamorphic.

7. Does the rock contain noticeable amounts of minerals other than quartz and feldspar, minerals such as garnet, kyanite, or epidote? These minerals (and others) form only in metamorphic rocks.

8. Does the rock contain parallel bands of different minerals or visible alignment of needle-shaped minerals? If so, the rock is metamorphic.

9. Is the rock glassy? If so, it is igneous.

IGNEOUS ROCKS

In all igneous rocks the minerals have an interlocking texture, characteristic of minerals that crystallize from a liquid. Most igneous rocks contain at least 50% feldspar and are classified as shown in Figure 3.2. Certain minerals have a strong tendency to group together as the molten rock (magma) crystallizes. One very common association is quartz, orthoclase feldspar, and biotite mica. A rock formed of these minerals is called granite, if the crystals are coarse enough to be visible, and rhyolite if the crystals are too small to see. Another common association is plagioclase feldspar, augite, and sometimes olivine. A rock with this mineralogy is called gabbro, if coarse-grained, and basalt, if fine-grained.

Classifying Igneous Rocks

To classify an igneous rock you must identify both its texture and mineral composition.

Step 1. Is the rock glassy (vitreous luster, frothy appearance, conchoidal fracture) or crystalline?

Figure 3.2

Identifying an igneous rock.

Texture	Mineral Composition	Rock Name		Origin
Nonfrothy Glass	Noncrystalline (glass, no minerals)	Obsidian	- - - - -	Extrusive
Frothy Glass	Noncrystalline (glass, no minerals)	Pumice	Scoria	Extrusive
Fine-Grained (minerals not visible)	Orthoclase feldspar, Quartz, Mica or Hornblende	Rhyolite		Extrusive
	Plagioclase feldspar, Augite or Hornblende		Basalt	Extrusive
Coarse-Grained (minerals visible)	Orthoclase feldspar, Quartz, Mica	Granite		Intrusive
	Plagioclase feldspar, Augite or Hornblende		Gabbro	Intrusive

Color		Light	Dark

Step 2. If the rock is crystalline, are individual crystals visible to the naked eye? If not, is the rock color light or dark?

Step 3. If the rock is coarse-grained, identify both the abundant minerals (at least 20% of the rock) and the less-abundant ones (accessory minerals).

Step 4. Place the rock in the appropriate pigeonhole and give it a name, for example, biotite granite or hornblende gabbro.

Some of the more common igneous rocks are illustrated on pages 20–21.

SEDIMENTARY ROCKS

Sedimentary rocks are of great importance to environmental scientists for several reasons.

1. They cover two thirds of the earth's land surface. Therefore, the characteristics of sedimentary rocks are critically important in most construction projects.

2. They are the host rocks for many economically important minerals, such as carnotite (uranium) for nuclear power plants, gypsum for drywall paneling in home construction, sulfur for the manufacture of sulfuric acid, and halite (table salt).

3. Almost all water, oil, and natural gas occurs in sedimentary rocks.

Sedimentary rocks can have either a fragmental or an interlocking texture (Figure 3.3). Those with an interlocking texture have crystallized from a liquid (water) and include chert, evaporites, and many limestones and dolostones. Fragmental sedimentary rocks are called conglomerates,

Obsidian glass, used by preindustrial peoples for making cutting tools and weapons.

Pumice, showing characteristic frothy vesicular texture. This glassy rock is used as an abrasive in Lava soap.

Scoria. Dark, vesicular rock formed as a glassy crust on basaltic lava flows.

Rhyolite. The crystals are too small to be visible to the unaided eye, but the pale pinkish color suggests the dominance of pink feldspar and quartz.

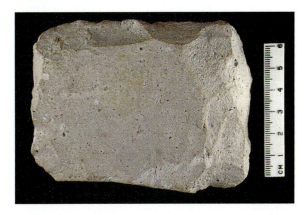

Basalt. The crystals are too small to be visible to the unaided eye, but the dark color suggests a large percentage of ferromagnesian minerals.

Granite. The pink mineral is orthoclase feldspar; the dark mineral is biotite mica; the clear grains are quartz; the white grains are plagioclase feldspar.

Gabbro. The dark mineral is augite; the light mineral is calcium-rich plagioclase.

sandstones, or mudrocks, depending on the size of the fragments (Table 3.1). Rocks that contain a mixture of sizes may be called muddy sandstone, sandy conglomerate, or other appropriate terms.

In geologic and environmental studies, it is also useful to describe the variation in sizes in fragmental rocks. This variation is called *sorting* (Figure 3.4). To a geologist, a well-sorted rock is one with a narrow range in grain size. To a construction engineer, however, a well-sorted rock is one with a wide range in grain sizes. The reason for this difference is that geologists are concerned with the efficiency of the natural processes that work to narrow the range of grain sizes, whereas construction engineers assess the value of grain aggregates for manufacturing concrete, in which a wide range of grain sizes is desirable. Hence, good sorting to a geologist is poor sorting to the concrete producer. Degree of sorting is also important in relation to groundwater flow; groundwater flows more easily (faster) through sandstones with a narrow range in grain size.

If a conglomerate or sandstone contains more than 90% quartz, it is called a quartz conglomerate or sandstone. If it contains less than 90% quartz, and feldspar is more abundant

Figure 3.3

Common textures of sedimentary rocks.

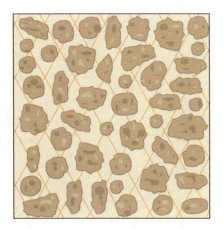

Fragmental texture: rounded or angular grains cemented by calcite.

Interlocking texture: crystals precipitated from solution.

TABLE 3.1

Grain-Size Scale for Sediment and Fragmental Sedimentary Rocks

Name of Rock	Name of Sediment	Name of Grain	Grain Size (mm)
Conglomerate	Gravel	Boulder	256–4,096
		Cobble	64–256
		Pebble	4–64
		Granule	2–4
Sandstone	Sand	Very coarse sand	1–2
		Coarse sand	0.5–1
		Medium sand	0.25–0.5
		Fine sand	0.125–0.25
		Very fine sand	0.062–0.125
Mudrock	Mud	Coarse silt	0.031–0.062
		Medium silt	0.016–0.031
		Fine silt	0.008–0.016
		Very fine silt	0.004–0.008
		Clay	< 0.004

↓

Figure 3.4

Sorting in sandstones.

Well sorted   Moderately sorted   Poorly sorted

than undisaggregated rock fragments, the rock is a feldspathic sandstone. If undisaggregated rock fragments are more abundant, the rock is a lithic conglomerate or sandstone (Figure 3.5).

Fragmental rocks composed of grains smaller than 0.06 mm form about 65% of sedimentary rocks and are called mudrocks because they form from compacted mud. They contain various proportions of silt and clay, grains too small to be seen with the naked eye. The average mudrock is mostly clay minerals, which are shaped like tiny micas, sheetlike with cleavage parallel to the sheets. In most mudrocks these clay minerals are well aligned and have a property called *fissility,* a tendency to split along very closely spaced parallel planes (Figure 3.6). This physical property gives mudrocks a strong tendency to slip past each other on slopes, a major cause of landslides. Fissile mudrocks are called shale.

Rocks composed entirely of calcite ($CaCO_3$) or dolomite [$CaMg(CO_3)_2$] form about 15% of sedimentary rocks (Figure 3.7). Those composed of calcite are called limestone; those composed of dolomite are dolostone. Limestone is more abundant than dolostone. Both rock types can either be fragmental or have the interlocking texture of a chemical precipitate. Those with interlocking texture are normally too fine-grained for individual crystals to be visible, much like a rhyolite or basalt. Fossils are common in both fragmental and nonfragmental carbonate rocks.

Figure 3.5a

Pebble conglomerate. Pebbles are mostly quartz and chert; sand matrix is mostly quartz. Cement is quartz.

(a)

Figure 3.5b–d

Coarse-grained sandstones of varying composition. (b) is all clastic quartz, cemented by quartz, so the rock is very light-colored; (c) is rich in feldspar (white grains) as well as quartz, and is cemented by hematite (reddish color); (d) contains abundant dark-colored rock fragments and is cemented by clay (grains too small to be visible). All three sandstones are fairly well sorted and may have some porosity. Most sand grains are about 1 mm in diameter; scale is in centimeters.

(b)

(c)

(d)

Calcite and dolomite are very soluble in water compared to nearly all other common minerals. Over periods of hundreds or a few thousand years, limestones and dolostones beneath the surface dissolve to form caves. Nearly all natural caves are simply giant holes dissolved in limestones. Sometimes the caves are so large and close to the surface that their roofs collapse, occasionally carrying buildings with them (Figure 3.8). Such collapse structures are common in limestone areas of the central Appalachians, Indiana, Missouri, Florida, and elsewhere.

Although sandstones, mudrocks, and carbonate rocks form at least 95% of sedimentary rocks, several other types are of sufficient environmental importance to deserve mention. Evaporites are deposits of very soluble minerals precipitated from pools of evaporating salty water. The most comon minerals in these deposits are halite (NaCl) and gypsum ($CaSO_4 \cdot 2\ H_2O$). Halite and gypsum are well known as table salt and plaster of paris.

Chert is a rock composed of microcrystalline quartz precipitated from solution. It occurs in many colors, including red (jasper), black (flint), and variegated (agate). Chert was used by preindustrial peoples for weapons and tools because of its hardness and splintery character (Figure 3.9).

Phosphorite is a rock composed of chemically precipitated crystals of the phosphate mineral apatite. It is a rich source of phosphate fertilizer and is mined extensively in western Wyoming and Idaho. Phosphorites also contain commercially important amounts of uranium and vanadium. The uranium is used as a fuel in nuclear power plants, the vanadium in specialty steels.

Taconite is the commercial name for banded iron ores found in Michigan, Minnesota, and several provinces of

Figure 3.6

Shale showing excellent fissility. The gray color results from the presence of about 1% organic matter.

Figure 3.7a

Limestone consisting of fossil shells about 2 cm in length set in a matrix of calcite crystals about 0.5 mm in diameter.

Figure 3.7b

Laminated microcrystalline limestone. Identified by its softness and the fact that it fizzes when acid is dropped on it.

Figure 3.8

One house down, one to go in Frostproof (but not sinkhole proof), Florida. The neighbor is making a hasty exit. (Note the moving van.)

Figure 3.9

Chert, a microcrystalline rock composed of SiO₂. Although it looks like microcrystalline limestone, chert is very hard and does not react in acid.

eastern Canada (Figure 3.10a). The bands are composed of alternating layers of chert and hematite. Taconite is the major source of America's iron for the manufacture of steel. A less-abundant but locally important type of iron ore is called ironstone and is formed of sand-size balls of hematite (Figure 3.10b).

Finally, there is coal, a sedimentary rock composed of land plant debris that has been subjected to deep burial, high pressure, and temperatures of 100–250° C. During burial, the plant material is chemically altered and its percentage of carbon is increased at the expense of other elements. The higher the carbon content, the better the coal. Coal use is increasing worldwide, as an easily mineable fuel and as a substitute for more expensive and harder-to-find petroleum. Unfortunately, the burning of coal causes more pollution than almost any other commonly used natural substance. Its only competitor for pollution honors is gasoline.

Figure 3.10a

Banded iron ore such as this is called taconite and is mined in the upper peninsula of Michigan. Dark bands are hematitic chert (jasper) of no economic value; light bands are a variety of hematite called specular hematite because its specks have a metallic appearance. Photo courtesy H. L. James.

Figure 3.10b

Rock formed of ooids of hematite (Fe_2O_3). The color is very distinctive, as is the reddish-brown streak.

Classifying Sedimentary Rocks

To classify a sedimentary rock, you must identify both its texture and mineral composition, as was the case for igneous rocks.

Step 1. If the rock is coarse-grained, is it fragmental? Fragmental rocks consist of rounded grains held together by chemically precipitated cement, most commonly calcite.

Step 2. If the rock is fragmental, what are the relative percentages of quartz, feldspar, and rock fragments? If it is a carbonate rock, does it dissolve when dilute acid is dropped on it? If it does, it is a limestone; if not, a dolostone.

Step 3. If the rock is fine grained, fragmental, and not composed of carbonate minerals, estimate its grain size. If it is microcrystalline, is it fissile? If microcrystalline and not fissile, is it hard (chert) or soft (mudrock)?

Step 4. Name the rock, for example, well-sorted, medium-grained sandstone, lithic pebble conglomerate, fossiliferous fragmental limestone, or black shale.

METAMORPHIC ROCKS

From the viewpoint of environmental science, the most important fact about metamorphic rocks is that most of them are *foliated,* a physical property similar in appearance to fissility in mudrocks. Slopes underlain by foliated metamorphic rocks can be as susceptible to landslides as shale slopes. What causes metamorphic rocks to be foliated?

Most sedimentary rocks are shales, and the abundant mineral in a shale is clay. The clay is transformed into mica at elevated temperature, an easy transition because both mineral types are sheetlike and have similar chemical compositions (Exercise 2). As burial temperature increases, shale changes into the metamorphic rock called slate, then into phyllite, then into schist (Figure 3.11a–c). Slate is much like shale except that its flat surfaces are more planar. In both shale and slate, the crystals of clay or mica are too

TABLE 3.2

Classification of the Common Foliated Metamorphic Rocks

	Crystal Size	Rock Names		Comments
Increasing Grade of Metamorphism ↓	Microscopic, very fine-grained	Slate (Figure 3.12a)		Well-developed planar surfaces but no sheen
	Fine- to medium-grained	Phyllite (Figure 3.12b)		Well-developed silky luster on mica surfaces
	Coarse-grained, mostly micas but often with other nonmicaceous minerals	Schist	Muscovite schist Chlorite schist Biotite schist Tourmaline schist Garnet schist (Figure 3.12c) Staurolite schist Kyanite schist Sillimanite schist Hornblende schist	Types of schist named on the basis of mineral content
	Coarse-grained, mostly nonmicaceous minerals	Gneiss (Figure 3.12d)		Well-developed color banding due to alternating layers of different minerals, most commonly quartz, feldspar, and dark-colored minerals

Quartzite (metamorphosed quartz sandstone) and marble (metamorphosed limestone or dolostone) are usually not foliated.

Figure 3.11a–g

a.–d. The important foliated metamorphic rocks. (a) Slate, identified by its aphanitic texture and very platy breakage pattern. (b) Phyllite, also aphanitic, but more micaceous, so a surface sheen of reflected light is present. (c) Schist, characterized by a predominance of coarse-grained micas. Dark red spots are garnet crystals. (d) Gneiss, composed of white feldspar, clear quartz, and dark streaks of biotite mica. Banding is diagnostic. (e) and (f) are equigranular quartzite and marble, respectively. They are the most common nonfoliated metamorphic rocks. (g) Natural asbestos, composed of three related fibrous minerals.

(d)

(a)

(e)

(b)

(f)

(c)

(g)

small to be seen. In a phyllite, the micas are coarser but still cannot be seen. They are, however, large enough to reflect light from their flat and parallel surfaces. Phyllites have a noticeable sheen in reflected light.

Probably, most metamorphic rocks are either slate, phyllite, or schist. Schist is composed of micas that are stable to temperatures of about 600° C, only 100° below the melting temperature of rocks. Above 600°, other, nonmicaceous minerals form. This new rock is called a gneiss (Figure 3.11d).

Micas are not the only minerals that form during metamorphism, but from an environmental point of view they are the most important. Other minerals in schists include tourmaline and garnet, semiprecious gemstones, and asbestos (Figure 3.11g), a cause of much concern in deteriorating buildings.

Classifying Metamorphic Rocks

Classifying a metamorphic rock requires recognizing its texture and mineral composition.

Step 1. Is the rock foliated or nonfoliated?

Step 2. If the rock is foliated, is it microcrystalline with very planar surfaces (slate), or microcrystalline with a sheen in reflected light (phyllite)? Does it have visible micas in parallel orientation (schist), or foliation defined by bands of different mineral composition (gneiss)?

Step 3. If the rock is foliated and coarse-grained, what are the abundant minerals?

Step 4. If the rock is not foliated, simply identify the abundant minerals.

Step 5. Name the rock. Typical names might be biotite-quartz schist, quartz-feldspar-garnet gneiss, or dolomite marble. Only the two or three most-abundant minerals are used in the rock name.

Problems

1. Describe and name each of the rocks in your tray.

2. What can you infer about the earth's history from the fact that a coarse-grained granite is exposed at Earth's surface?

3. Which type(s) of rock do you think is (are) likely to contain more pore space, a sandstone, a limestone, or a mudrock? Explain. What practical importance might this have?

4. Sands are usually deposited in a different geographic location than are clay minerals. What is the reason for this?

5. Micas and clay minerals can be compacted to a much greater degree than quartz and feldspar grains. Explain why. What is the practical importance of this fact?

6. Why are slopes underlain by shale always unstable?

7. You are choosing a tombstone to mark your final resting place. Would you have it carved from granite, quartzite, or marble? Explain. Suppose the climate were arid. Would this make halite a satisfactory rock for your tombstone? Explain.

8. Which type(s) of metamorphic rock would be the best choices for roadbed material?

9. One type of metamorphic rock has been widely used in place of wooden or composition roofing shingles. The same rock was used in the last century to make writing tablets for schoolchildren. Name the rock and explain why it is so suitable for these uses. Why do you think it is rarely used these days for either roofs or writing tablets?

10. What types of rock would be most durable for use as facing for buildings such as banks or major office buildings? Explain.

Further Reading/References

Blatt, Harvey, 1992. *Sedimentary Petrology,* 2nd ed. New York, W. H. Freeman, 514 pp.

Blatt, Harvey, and Tracy, Robert, 1996. *Petrology.* New York, W. H. Freeman, 529 pp.

Dietrich, Richard V., 1989. *Stones: Their Collection, Identification and Uses,* 2nd ed. Prescott, Arizona, Geoscience Press.

Hannibal, J. T., and Park, L. E., 1992. "A guide to selected sources of information on stone used for buildings, monuments, and works of art." *Journal of Geological Education,* v. 40, pp. 12–24.

Mack, Walter N., and Leistikow, Elizabeth A., 1996. "Sands of the world." *Scientific American,* August, pp. 62–67.

Igneous Rocks

Sample Number	Texture	Percent Quartz	Percent Orthoclase	Percent Plagioclase	Type and Percent Accessories	Name of Rock

Sedimentary Rocks

Sample Number	Clastic or Nonclastic	Crystal or Grain Size	Mineral Composition	Cement if Clastic	Rock Name

Metamorphic Rocks

Sample Number	Foliated?	Major Minerals	Name of Rock

4 EXERCISE

TOPOGRAPHIC MAPS

A map is a representation, usually on a flat surface, of a part of the earth's surface. It can be an actual photograph taken from an orbiting satellite or an airplane, or a numerical approximation of the surface, with lines marking certain boundaries or elevations. Some maps show elevations above sea level (topographic maps; Figure 4.1) or below sea level (bathymetric maps). Some show the contacts between rock units of geologic significance (geologic maps) or agricultural significance (soil maps). Still others are derivative maps constructed for environmental or engineering purposes—for example, flood-frequency maps, land-use maps, and maps of groundwater composition. In this and subsequent exercises we will consider several types of maps that are essential tools for environmental geologists.

MAP SCALE

Features on a map are smaller than the actual features they represent. This reduction in size is termed the *scale* of the map. A scale of 1:100 means that one unit on the map is equal to 100 of the same units on the earth's surface. In geological studies, a common scale for maps is 1:62,500, meaning that one inch on the map equals 62,500 inches on the ground—approximately one mile (1 mi = 63,360 in.). A scale of 1:125,000 is a *smaller* scale than 1:62,500 because an inch represents two miles on the ground rather than one mile. A feature on the ground must be larger to be visible on this smaller-scale map. Many maps used in environmental studies have a numerical scale of 1:24,000 (one in. = 2,000 ft). In addition to a numerical scale, most maps have a graphic scale, a line or bar divided into segments that represent units of length on the ground.

MAGNETIC DECLINATION

The earth rotates around an imaginary line that passes through the earth's center. The sites where this line intersects the surface of the earth are called the *north and south geographic poles,* with geographic north shown on maps by an arrow. Also shown on geologic and many other types of maps is magnetic north, the direction to the *magnetic north pole.* Magnetic directions arise from the fact that the earth acts like a simple bar magnet, with the imaginary bar passing through the center of the earth and intersecting the surface near, but not at, the geographic poles. Many maps have arrows pointing to both the geographic and magnetic poles; the angle between them is the *magnetic declination.* When we use a magnetic compass to determine directions in the field, we adjust it to compensate for the magnetic declination, so that the needle points to "true" (geographic) north.

MAP COORDINATES

Maps of use to environmental geologists range in scale from very small to very large. For example, a map showing the outcrop pattern of America's premier underground water source, a rock unit known as the Ogalalla Formation, needs to be small scale because the Ogalalla extends over five midwestern states. The map needs to cover a very large area. In contrast, a map showing the location of a city's sewer lines that feed into the sewage treatment plant needs to be at a much larger scale.

Locations on small-scale maps are usually given by a worldwide system of coordinates called *latitude* and *longitude* (Figure 4.2). Latitude describes how far north or south of the

Figure 4.1

Reduced copy of the Baraboo, Wisconsin, 15-minute quadrangle, with principal features highlighted. *Source: U.S. Geological Survey.*

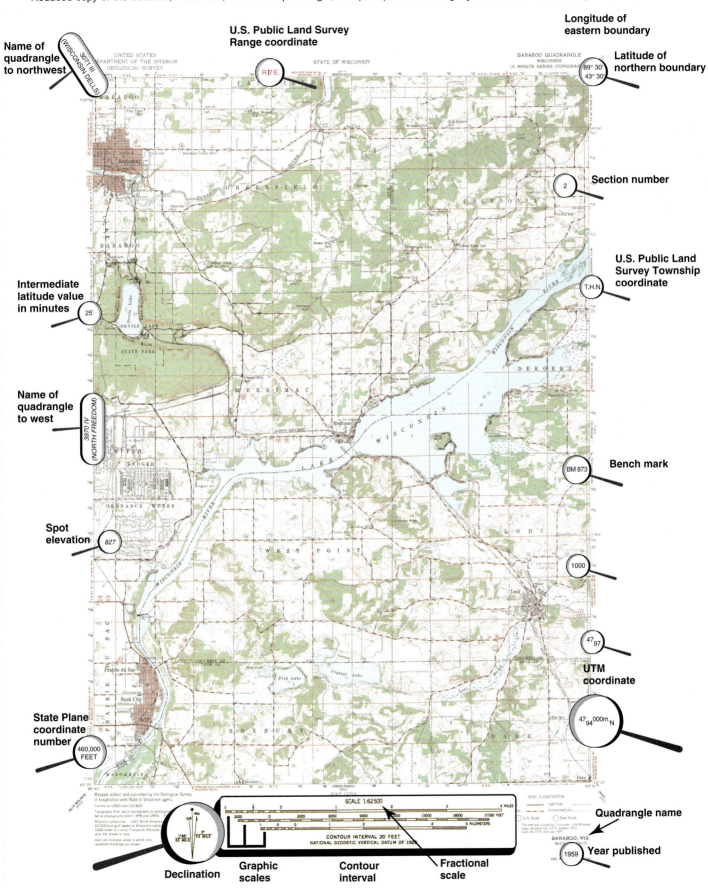

Name of quadrangle to northwest — 3071 II (WISCONSIN DELLS)

U.S. Public Land Survey Range coordinate — R.7E.

Longitude of eastern boundary — 89° 30'

Latitude of northern boundary — 43° 30'

Section number — 2

U.S. Public Land Survey Township coordinate — T.H.N.

Intermediate latitude value in minutes — 25'

Name of quadrangle to west — 3970 IV (NORTH FREEDOM)

Bench mark — BM 873

Spot elevation — 827

1000

UTM coordinate — 47 97

State Plane coordinate number — 460,000 FEET

47 94 000m.N

SCALE 1:62500

CONTOUR INTERVAL 20 FEET
NATIONAL GEODETIC VERTICAL DATUM OF 1929

Declination

Graphic scales

Contour interval

Fractional scale

Quadrangle name

BARABOO, WIS.

Year published — 1959

Figure 4.2

The basis of numbering latitude and longitude.

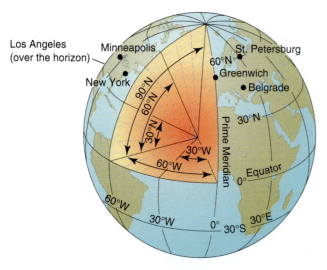

equator a place is and is measured in degrees. The equator is 0° of latitude. A place at 30° latitude lies where a line from the surface to the earth's center intersects the plane of the equator at an angle of 30°. Latitude lines on the surface are parallel to each other. Both Phoenix, Arizona, and Baghdad, Iraq, are at 33° north latitude.

East-west locations on the earth's surface are given by longitude. Zero longitude is a north-south line from pole to pole that runs through Greenwich, England, as established in 1884 when England was the premier world power. It is called the Prime Meridian. A place at 30° longitude lies where a line from the surface to the earth's center intersects the plane of the Prime Meridian at an angle of 30°. Meridians (longitude lines) are numbered both east and west from the zero line at Greenwich from zero to 180° east and zero to 180° west. The 180° line is a meridian that runs through the western Pacific Ocean. All locations on a north-south line have the same longitude. The area of the 48 conterminous United States is roughly shaped like a rectangular box bounded by 38° and 48° N latitudes and by 70° and 124° W longitudes.

In 1785, the federal government instituted the *Public Land Survey System* to deal with the problem of geographic locations in small areas during land surveying for westward expansion. Thirty-four of the fifty states are subdivided according to the PLSS. Most of the other 16 were part of the original 13 colonies.

The PLSS was established in each state by surveying at least one east-west *base line* and one north-south *principal meridian* (Figure 4.3). Adjacent states sometimes had the same base lines or principal meridians. Once the lines were established and related to latitude and longitude, additional lines parallel to them were drawn at six-mile intervals throughout each state, creating grids of squares, each square six miles on a side.

Squares along each east-west strip of the grid were called *tiers* or *townships* and were numbered, relative to the base line, as Township 1N, T2N, T3N (or T1S, T2S), etc. Squares along each north-south strip were called *ranges* and were numbered relative to the principal meridian (R1E, R2E, etc.).

Figure 4.3a–c

Generalized diagram of the PLSS. (a) Tier and range grid. (b) One township, 6 × 6 miles. (c) One section, 1 × 1 mile.

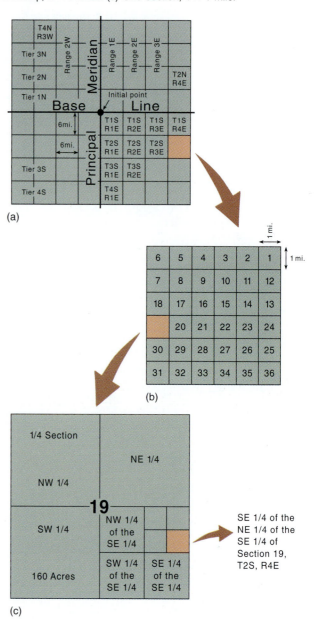

Each of the six-mile squares (which, unfortunately, was also called a township) was further subdivided into 36 one-mile squares called *sections* and numbered in the rather peculiar pattern shown in Figure 4.3b. In flat areas of the mid-continental United States, most of the main rural roads follow section lines. Subdivision of sections (Figure 4.3c) was less formal and introduced no new terms. Reference might be made to the S 1/2 of a section, or perhaps to the SE 1/4 of a section. More specific locations were designated simply as quarters of quarter-sections—for example, the NW 1/4 of the NE 1/4 (an area of 40 acres)—and so on.

ELEVATIONS

Elevations on maps are ultimately referenced to mean sea level as determined very accurately by the United States

Coast and Geodetic Survey. More local reference points are *bench marks*, inland elevations determined accurately by the United States Geological Survey. Bench marks are widely scattered, so a particular map area might not contain one. The map might, however, include points whose elevation above sea level has been fairly accurately determined.

CONTOUR LINES

Contour lines are lines that connect points of equal value of a variable. Thus, a topographic contour line connects points of equal elevation above sea level; an *isohyet*, points of equal precipitation; an *isopach*, points of equal thickness of a rock unit; an *isotherm*, points of equal temperature; and an *isoseismal line*, points of equal earthquake intensity. The difference in value between adjacent lines is called the *contour interval*. The choice of contour interval is arbitrary and based on the scale of variation in the property being considered. For example, a map of an area in Kansas might require a topographic contour interval of 10 feet to show important features; in neighboring Colorado the contour interval might vary from 10 feet in the eastern part of the state to 100 feet in the western, more mountainous part.

TOPOGRAPHIC MAPS

Topographic maps show the shape of the earth's surface by means of contour lines. Variations in the surface shape help control streamflow rates, the likelihood of landslides, the probability of flooding, and many other things important to environmental geologists. A topographic contour line, which connects points of equal elevation above sea level, is formed by the intersection of an imaginary level surface with the ground (Figure 4.4). A natural example of a contour line is a shoreline around a lake. An example created by people of successive contours at different elevations is the pattern produced by contour plowing, a plowing pattern designed to reduce soil erosion. Contour lines that cross a low area such as a stream channel will form a V-shape that points upstream (Figure 4.5).

Figure 4.4

Properties of Contour Lines

> **Properties of Contour Lines***
>
> 1. Every point on a *contour line* is the exact same elevation; that is, contour lines connect points of equal elevation.
>
> 2. Contour lines always separate points of higher elevation (uphill) from points of lower elevation (downhill). One must determine which direction on the map is higher and which is lower, relative to the contour line in question, by checking adjacent elevations.
>
> 3. The elevation between any two adjacent contour lines on a topographic map is the *contour interval*. Often every fifth contour line is heavier, so you can count by five-times the contour interval. These heavier contour lines are known as *index contours*, because they generally have elevations printed on them.
>
> 4. Contour lines never cross one another, except in one rare case: an overhanging cliff. In such a case, the hidden contours are dashed.
>
> 5. Contour lines merge to form a single contour line only where there is a vertical cliff.
>
> 6. Contour lines never split.
>
> 7. Contour lines that are
>   a. evenly spaced indicate comparatively uniform slopes.
>   b. closely spaced show steep slopes.
>   c. widely spaced portray gentle slopes.
>   d. irregularly spaced signify irregular slopes.
>
> 8. Contour lines form a V pattern when crossing streams. The apex of the V always points upstream (uphill).
>
> 9. A concentric series of closed contours represents a hill:
>
> 10. *Depression contours* have hachure marks on the downhill side, always close, and represent a closed depression:
>
>   a. Where the topography slopes downhill and a standard contour is adjacent to a hachured contour, the hachured contour is one contour interval lower than the standard contour.
>   b. Where the topography slopes uphill and a standard contour is adjacent to a hachured contour, the two contours have the same elevation.
>   c. Where one closed hachured contour encloses another closed hachured contour, the inner contour is one contour interval lower.
>   d. Where a closed standard contour is enclosed by a hachured contour, they both have the same elevation.
>
> *These same rules apply to bathymetric maps, which show the topography of the floors of lakes or oceans.

Figure 4.5

The relationship between the shape of the land surface and contour lines on a topographic map. Steep slopes are shown by closely spaced contours, gentle slopes by widely spaced ones. In this example, the steep slope results from resistant layered rocks and the gentler slope from less resistant rocks such as shales. Note that when contours cross streams, they *V* upstream. The sand spit at the bottom of the map has no elevation that reaches 20 feet above sea level, so that no contour lines are present on the spit. *Source: U.S. Geological Survey.*

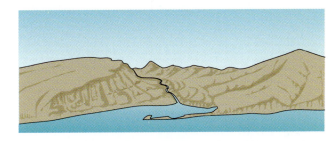

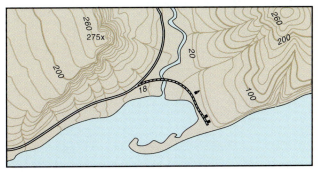

Most precise locations do not fall exactly on a contour line—for example, a spring might be located between contour lines representing 430 and 440 feet above sea level. In such cases, we estimate the elevation by measuring with a graduated scale from the spring to one of the contour lines. If the spring is 65% of the distance from line 430 to line 440, for example, its elevation is approximately 436.5 feet. The elevation can only be approximated because this method assumes that the ground slopes evenly between the two contour lines. In reality, the surface might slope gently from 430 to 433 feet, then more steeply from 433 to 440 feet. In such a case, the method would yield a slightly erroneous, higher elevation for the spring. The accuracy of the estimate, therefore, depends on both the contour interval and the ground slope. A five-foot contour interval, rather than ten-foot, would reduce the potential for error.

Relief is defined as the difference in elevation between two points on a map. *Local relief* is the maximum difference in elevation within a designated small area on the map; *total relief* is the maximum difference in elevation between any two points on the map. Local relief helps control stream gradient and the frequency of large-scale earth movements such as landslides, and it may also reflect relatively recent movement along breaks (faults) in the earth's crust.

TOPOGRAPHIC PROFILES

A topographic map provides a view from above the ground surface, using symbols and contour lines to show physical

Figure 4.6

Topographic profile across the Pseudomountain area along line A-A'. Vertical exaggeration = 32×. *Source (Top): U.S. Geological Survey.*

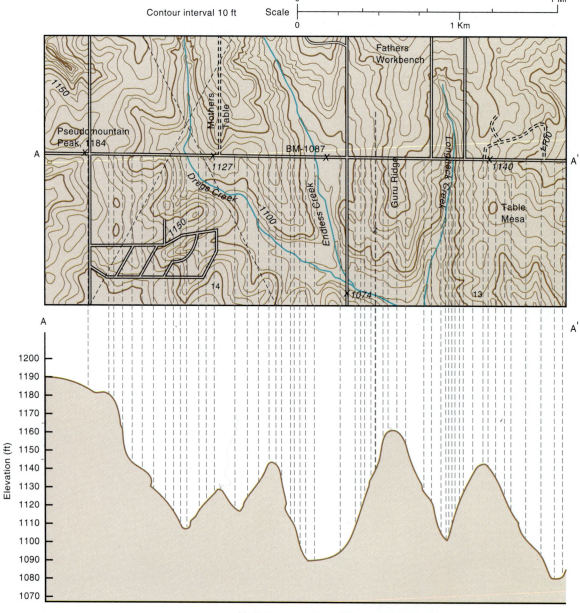

Vertical exaggeration = 32x.

features and relief. This depiction permits us to visualize the shape of the ground surface and the location of buildings, rivers, and other features. A *topographic profile* is a cross-section that shows the changing elevation of the land surface along a line. We select the position of the line either to provide a representative sample of the surface irregularities shown on the map or to show the configuration of the ground surface in an area of particular interest. Figure 4.6 illustrates the construction of a topographic profile (or cross-section). The procedure is as follows:

1. Draw the line along which you wish to construct the profile. Note the maximum topographic relief along the line.

2. Choose a vertical scale. Scales are arbitrary, but most have about 10 to 15 divisions between the lowest and highest elevations on the cross-section. Using fewer than 10 divisions risks losing desired details of the ground surface; using more than 15 adds more detail than is needed for most purposes. Choose a number of divisions that suits the intended purpose of the profile. Label the horizontal lines of the profile grid.

3. If the profile line runs exactly east-west, simply drop a dashed line from each point at which the profile intersects a contour line to the appropriate elevation on the profile grid, as shown in the illustration. Connect the dots with a smooth curve.

4. If the profile line does not run exactly east-west, place the upper edge of a clean strip of paper along the line. Where each contour line intersects the profile line, make a short tick mark on the paper and write the elevation by it. Then place the paper along the bottom edge of the profile grid and project from each tick mark up to the correct grid elevation.

Just as a map is always smaller than the area it represents, on any cross-section the vertical distance between two elevations on the paper is less than the true vertical distance (difference in elevation) on the ground. The vertical scale nearly always differs from the horizontal scale. The ratio between the fractional vertical scale and the fractional horizontal scale is called the *vertical exaggeration* (Figure 4.7). For example, if the horizontal scale is 1:62,500 and the vertical scale is 1:1,200, the vertical exaggeration is approximately $52\times$ (62,500/1,200). This means that the slopes on the topographic cross-section are 52 times steeper than the corresponding slopes on the ground.

CONSTRUCTING A TOPOGRAPHIC MAP

Many times an environmental geologist must create a contour map from scattered known elevations. For example, a few bench marks may be present together with a few points whose elevations are known from earlier surveys made when the town was established. In drawing contour lines from such scattered data, there are a few general guidelines.

1. Note the difference in elevation between the lowest and highest points in the area of interest to decide on an appropriate contour interval. The greater the range in elevations, the wider the contour interval is likely to be. However, the appropriate contour interval also depends on how many data points exist and how much detail they provide. For example, if all the elevation points you start with differ from each other by more than 100 feet, you cannot make a reliable map with a 10-foot contour interval. You would be guessing at the position of most contour lines.

2. Note the location of low and high points. The starting data points will help, as will the flow directions of any streams in the area if you know them. Water always flows downhill. Also note where slopes seem to be steep, places where your data points are close together and differ greatly in value.

3. Start contouring from the lowest elevations on the map. Remember that contours always V toward

Figure 4.7

Distortion in a profile of a human face caused by different vertical exaggerations.

Exaggerated 5 times

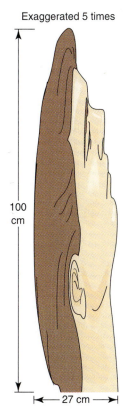

100 cm

27 cm

Exaggerated 2½ times

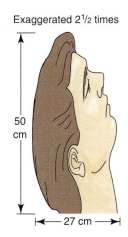

50 cm

27 cm

No exaggeration

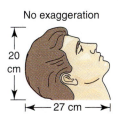

20 cm

27 cm

higher elevations when crossing low areas such as valleys. Label all contour lines as you draw them so you can keep track of where you are headed and will draw lines that are consistent with the data points you started with.

4. Avoid constructing imaginary slopes by forcing contour lines to run down the side of the map. It will probably be more accurate to run the line off the map at one place and have it enter again from another place. Sometimes the contour lines are properly drawn parallel to the border of the map, but be sure that your lines are consistent with the elevation points you started with.

Problems

1. Examine Figure 4.6.
   a. What are the highest and lowest elevations on the map?
   b. Is Mothers Table higher than Fathers Workbench? What is the difference in feet between them?
   c. If Pseudomountain Peak is in section 18, what are the numbers of the other sections on the map?
   d. Which way do the streams flow? How do you know?
   e. What is the gradient of Longneck Creek?
   f. What is the scale of the map?
   g. An acre is 43,560 square feet. How many acres are on one section of land?

2. Make a topographic map by contouring the following points, using a 50-foot contour interval. Use a pencil, and sketch lightly so you can erase errors easily. Numbers indicate elevation in feet above sea level for each adjacent dot. Use the contour provided on the map as a starting guide.

3. Shown is a portion of the Waite, Maine, 7 1/2′ Topographic Quadrangle. Based on this map, answer the following questions.
   a. How many square miles are shown on the map?
   b. What is the straight-line distance from Waite to Bingo?
   c. Why do you think the highway between the two towns was not built along the straight line?

Map for problem 2.

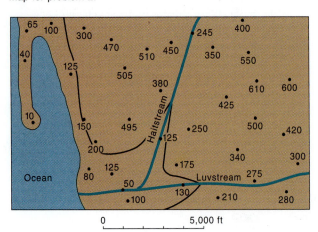

Map for problem 3.

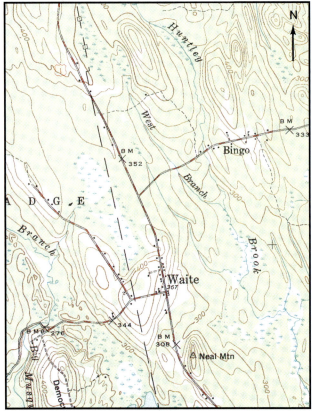

1:24,000

d. What is the total relief on the map?
e. In which direction does Huntley Brook flow?
f. What is the relationship between the direction in which streams intersect on the map and the direction in which they flow?
g. Construct a topographic cross-section from BM 276 to BM 333. What is the vertical exaggeration of your cross-section?

Further Reading/References

Bart, H. A., 1991. "A hands-on approach to understanding topographic maps and their construction." *Journal of Geological Education,* v. 39, pp. 303–305.

Miller, V., and Westerback, M. E., 1988. *Interpretation of Topographic Maps.* Columbus, Ohio, Merrill, 416 pp.

Muehrcke, P., and Muehrcke, J.O., 1992. *Map Use: Reading Analysis and Interpretation,* 3rd ed. Madison, Wisconsin, JP Publications.

Thompson, M. M., 1981. *Maps for America,* 2nd ed. Reston, Virginia, U.S. Geological Survey, 265 pp.

Upton, W. B., Jr., 1970. *Landforms and Topographic Maps.* New York, John Wiley & Sons, 134 pp.

E X E R C I S E 5

GEOLOGIC MAPS

Topographic maps show the shape of the ground surface. *Geologic maps,* in contrast, show the areal distribution and orientation of the three-dimensional bodies of rock that appear on the surface. These surface appearances are termed *outcrops.* The entire Earth is, of course, underlain by rocks, but they are not always evident. Many humid tropical or subtropical areas, such as central Africa, the Amazon region of South America, and the southeastern United States, have few outcrops because their soils are so thick—sometimes tens of feet. In the northeastern and north-central United States and Canada, outcrops are scarce because they are covered by Pleistocene glacial debris. In contrast, outcrops are often both extensive and continuous in arid or semiarid regions such as those of the Middle East and western Texas. Where outcrops are scarce, geologists making a map infer probable rock distributions from surrounding areas with better outcrops. Thus, geologic maps are not maps of actual outcrops; rather, they incorporate both actual outcrops and geologic inference. A location shown as sandstone on a geologic map might in fact be a wheat field; we would find the sandstone only by digging through the soil.

Geologic maps show numerous features relating to the origin and geologic history of an area, many of them relevant to environmental concerns. Environmental geologists, therefore, must know how to read and interpret geologic maps. Geologic maps might suggest the best sites for placing sanitary landfills, drilling water wells, finding springs, or finding construction-grade sand and gravel, as well as reveal areas most vulnerable to landslides or earthquakes, or many other things of environmental importance. Special-purpose maps derived from geologic maps emphasize specific geologic hazards for engineering or environmental analyses.

What features related to rock distributions do geologic maps show (Figures 5.1, 5.2)?

1. *Formations.* Geologists define a formation as a mappable rock unit defined by rock type (e.g., sandstone, shale), geologic age, fossil content, or some other characteristic that is both easy to see and diagnostic. Formations are given names, such as the Wellington Formation or Redwall Limestone, and assigned colors and symbols to identify them on a map. The Wellington Formation, which is of Permian age, might be assigned the symbol "Pw" (Permian, Wellington). Redwall Limestone might be "Mr" (Mississippian, Redwall). Different formations are shown on a map by colors or line patterns.

2. *Folds.* A fold is a bend in a layered rock. An "upfold" is called an *anticline;* a "downfold" is a *syncline* (Figure 5.3). In an eroded anticline the oldest layers are in the center of the fold, while in an eroded syncline the youngest layers are in the center.

3. *Faults.* A fault is a break in the earth's crust along which movement has occurred (Figure 5.4). The rocks on one side of a fault might move up, down, or sideways with respect to the other side, but the movement is always parallel to the fault surface. Paved areas might crack, but only in movies does the earth pull apart, perpendicular to the fault, and swallow whole towns.

4. *Contacts.* A contact is the boundary between two rock units or between a rock unit and a fault.

Figure 5.1

Geologic symbols for rock types. *Source: U.S. Geological Survey.*

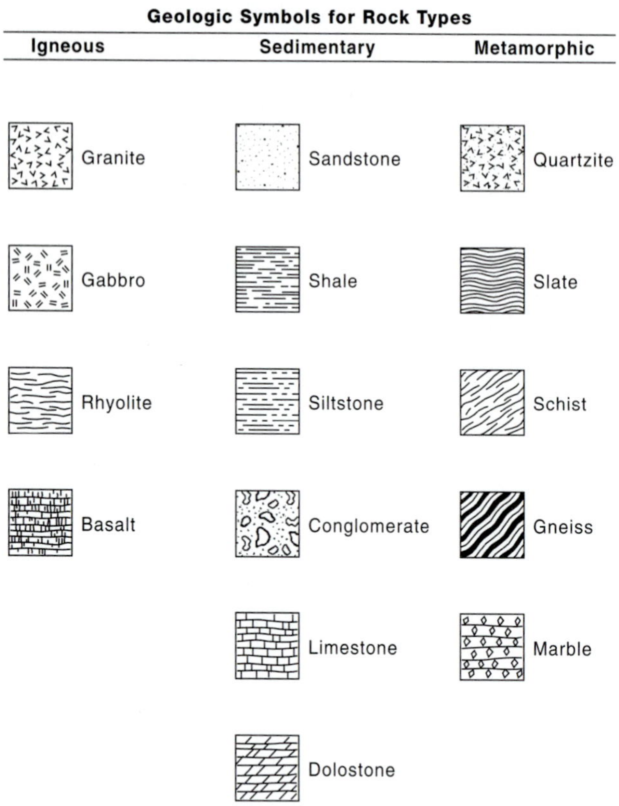

Geologic Symbols for Rock Types

Igneous	Sedimentary	Metamorphic
Granite	Sandstone	Quartzite
Gabbro	Shale	Slate
Rhyolite	Siltstone	Schist
Basalt	Conglomerate	Gneiss
	Limestone	Marble
	Dolostone	

5. *Strike.* The strike of a tabular rock layer is the compass direction of the line formed by the intersection of the layer with an imaginary horizontal plane (Figure 5.5).

6. *Dip.* The dip of a rock layer is the vertical angle formed by the intersection of the layer with an imaginary horizontal plane (Figure 5.5). The dip angle must be measured perpendicular to the strike direction; when measuring in any other direction the angle determined will be less than the true dip.

7. *Unconformities.* An unconformity is an ancient erosional surface covered by later sediment. The present ground surface is an unconformity in the making. Unconformities are normally identified in the map legend and may not be evident on the map itself.

8. *Other features.* A geologic map may also include some of the features shown on topographic maps, such as topographic or bathymetric contours, streams, or cities.

ENVIRONMENTAL GEOLOGY APPLICATIONS

Geologists use the location and geologic age of map features to unravel the geologic history of the map area. Similarly, numerous features on geologic maps are important in environmental geology.

1. The *lithology* of a formation helps determine its ability to transmit such fluids as water, petroleum, and natural gas. A sand deposit, for example, has spaces between the grains that can fill with fluids, while a chemical precipitate such as gypsum does not.

2. Shale and clay deposits are rich in clay minerals. The platy shape of these minerals causes them to serve as slip surfaces on which landslides begin.

3. Faults often permit the movement of fluid through rocks that might otherwise be impermeable. The same is true of *joints,* parallel rock fractures along which no movement has occurred. Unconformity surfaces can also serve as fluid conduits.

4. The angle of dip of a fluid-containing layer helps determine the rate at which the fluid flows through the layer. Water flowing quickly through a rock can mean a more readily available supply for a nearby city.

Geologic maps are an essential tool for the environmental geologist. When combined with topographic maps and an understanding of rocks and minerals, they form the basis for making important decisions on environmental issues.

GEOLOGIC CROSS-SECTIONS

A geologic map shows the locations of all the major rock units present on the ground surface, or that *would be* present if soil cover, glacial debris, and human constructions were removed. But what happens below the ground surface? Does a water-bearing sandstone gradually become finer-grained, grade into shale, and lose its water-bearing properties? What is the source of water leaking from a fault surface as a spring? Do limestones exposed at the surface become cavernous underground, posing a danger to buildings constructed directly above? These are some of the reasons why environmental geologists must be able to visualize the materials buried beneath the ground surface. The result of this visualization is a *geologic cross-section.*

Some of the data used in constructing a cross-section, such as strike and dip, come from observations made at ground level. Others are obtained from wells drilled to find water, petroleum, and natural gas; from mines excavated for mineral resources; or from tunnels excavated to enlarge transportation networks. Wells drilled for the petroleum industry are useful because they bring rock chips to the surface, and because they are "logged" in some form. Wireline logs are zigzag lines on a strip of paper that indicate how subsurface rocks respond to electrical or nuclear impulses. The character of the response allows geologists to interpret lithology, water content, and other useful properties of the rock. Another technique for analyzing subsurface rocks is to use explosives to send energy impulses into the ground. Contacts between rock layers reflect some of the energy upward, permitting determination of the type of rock and its

Figure 5.2

Symbols on geological maps published by the United States Geological Survey. *Source: U.S. Geological Survey.*

Strike and dip of strata	
Vertical strata	
Strike and dip of overturned strata	
Horizontal strata	
Strike and dip of rock cleavage	
Vertical cleavage	
Axis of an anticline (concave-downward folded rock)	
Axis of a syncline (concave-upward folded rock)	
Axis of a plunging anticline (plunging means the crest of the fold is not horizontal)	
Axis of a plunging syncline	
Trend and angle of plunge of a line	
Lateral (strike-slip) fault; arrows indicate relative movement	
High-angle fault; U for Up and D for Down, to indicate relative movement	
Reverse fault; Teeth are in the side of the hanging wall (upper block)	
Contact or other line solid where known, dashed where approximated, and dotted where only inferred	

Shafts:
Vertical   Inclined

Adit, tunnel, or slope:
Accessible   Inaccessible

Prospect

Quarry:
Active   Abandoned

Gravel pit:
Active   Abandoned

Oil well:
Drilling   Shut-in   Dry hole abandoned
Gas   Show of gas
Oil   Show of oil

Q	Quaternary
T	Tertiary
K	Cretaceous
J	Jurassic
Ŧ	Triassic
P	Permian
ℙ	Pennsylvanian
M	Mississippian
D	Devonian
S	Silurian
O	Ordovician
Є	Cambrian
P-Є	Precambrian

dip. The dip of the layer at the surface is not necessarily the same as its dip below the surface.

Constructing a geologic cross-section is very much like constructing a topographic profile:

1. Orient the cross-section line approximately normal to the major geologic trends in the area, that is, normal to the strike of beds, folds, and faults.

2. Construct a topographic profile along the cross-section line.

3. Use tick marks to transfer the geologic contacts along the line to the topographic profile, as was done for elevations in the topographic profile.

4. Where dips are known, extend formational contacts and fault planes down into the profile, using a protractor to measure angles.

5. Connect at depth all outcrops of the same sedimentary rock unit, using a reasonable extension of the surface dips. Most units maintain a fairly constant thickness. If the cross-section includes a fault, be aware that rock units will not match up exactly across the fault. If one side of the fault has moved up or down, the formation contacts will be displaced across the fault surface.

6. Label the cross-section, using the appropriate letter symbols from the map. Also place arrows alongside any faults to show relative displacement.

Figure 5.3

Anticlines fold upward; synclines fold downward.

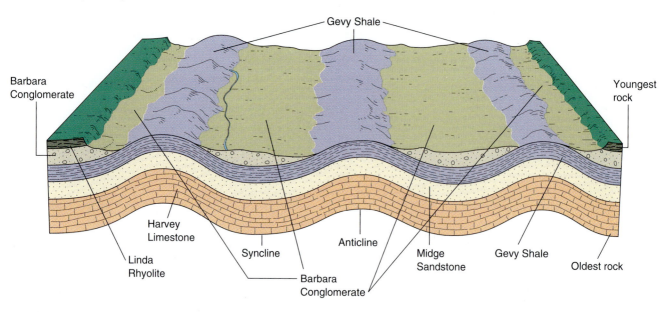

Figure 5.4

Four types of faults are illustrated in these block diagrams. If the fault plane is inclined, the side above the fault is the hanging wall and the side below is the footwall. Displaced marker beds show movement.

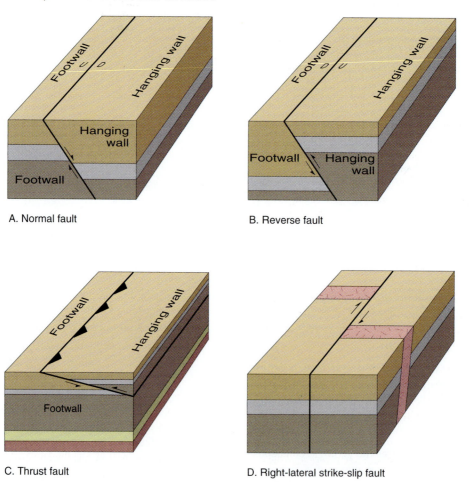

A. Normal fault

B. Reverse fault

C. Thrust fault

D. Right-lateral strike-slip fault

Figure 5.5

Strike and dip of a limestone layer resting on granite. Strike is east-west and dip is to the south at about 30°.

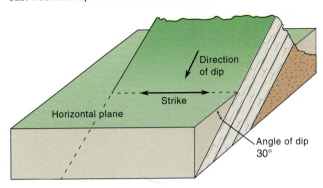

PRINCIPLES OF GEOLOGIC MAP INTERPRETATION

After a geologic cross-section is constructed, it must be interpreted to answer questions such as the relative ages of the rock layers and faults or the ages of unconformities. The geologic principles on which these interpretations are made are the following:

1. *Original horizontality.* Layers of sedimentary rock were originally deposited as horizontal layers of sediment. If these rocks are now dipping, they must have been tilted after deposition.

2. *Superposition.* In an undisturbed sequence of sedimentary rocks, the oldest rock is at the bottom. Each succeeding layer is younger than the layer below it.

3. *Cross-cutting relationships.* A feature that cuts across a layer of sedimentary rock must be younger than the layer being cut. Thus, a fault is younger than the beds it displaces, and an igneous intrusion is younger than the rocks it intrudes. An unconformity must be younger than the rocks it cuts. A layer must be present before it can be cut, intruded, or eroded.

4. *Inclusions.* An inclusion is a fragment of a pre-existing body of rock, present within a younger rock. For example, a magmatic intrusion might contain a piece of sedimentary rock broken off during the ascent of the magma. The intrusion must be younger than the piece of sedimentary rock. Commonly, the sedimentary rock layer directly above an unconformity surface contains fragments from rock layers below the unconformity. An unconformity is an ancient erosional surface—an ancient ground surface—and rock fragments were lying on the surface when the sedimentary layer above the unconformity was deposited. Inclusions, although often visible in outcrops, are not normally indicated on geologic maps.

AN EXAMPLE OF GEOLOGIC INTERPRETATION

Figure 5.6 is a block diagram showing in three dimensions the relationships among some igneous and sedimentary rocks. Using the four principles listed above, we can determine the sequence of events that resulted in the picture we see.

1. The oldest rock we can see is dolostone at the base of the series of sedimentary rocks.

2. After deposition of the dolostone, the sandstone was deposited.

3. The dolostone and sandstone were then tilted toward the east, and the parts of these rocks that protruded above the surrounding countryside were eroded to form a flat surface.

4. Limestone was deposited on top of the dolostone and sandstone. The contact between the limestone and these underlying rocks is an unconformity (labeled 1).

5. Shale was deposited above the limestone.

6. Granite intruded through the dolostone, would have gone through the sandstone if it were present at the site of the intrusion, and entered the limestone. It is possible that the intrusion came after deposition of the shale, but there is no way to tell from the rocks as we see them. The granite contains inclusions only of the dolostone it penetrated.

7. A tabular body of rhyolite intruded through the dolostone, sandstone, limestone, shale, and whatever lay above the shale, if anything. If there were nothing above the shale, the magma would have spilled onto the shale ground surface as a lava flow.

8. The rhyolite, and probably some shale, were eroded to form a flat surface.

Figure 5.6

Block diagram revealing geologic history.

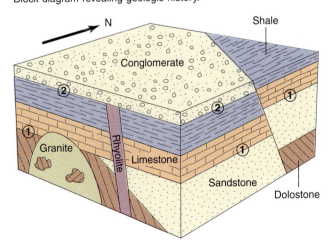

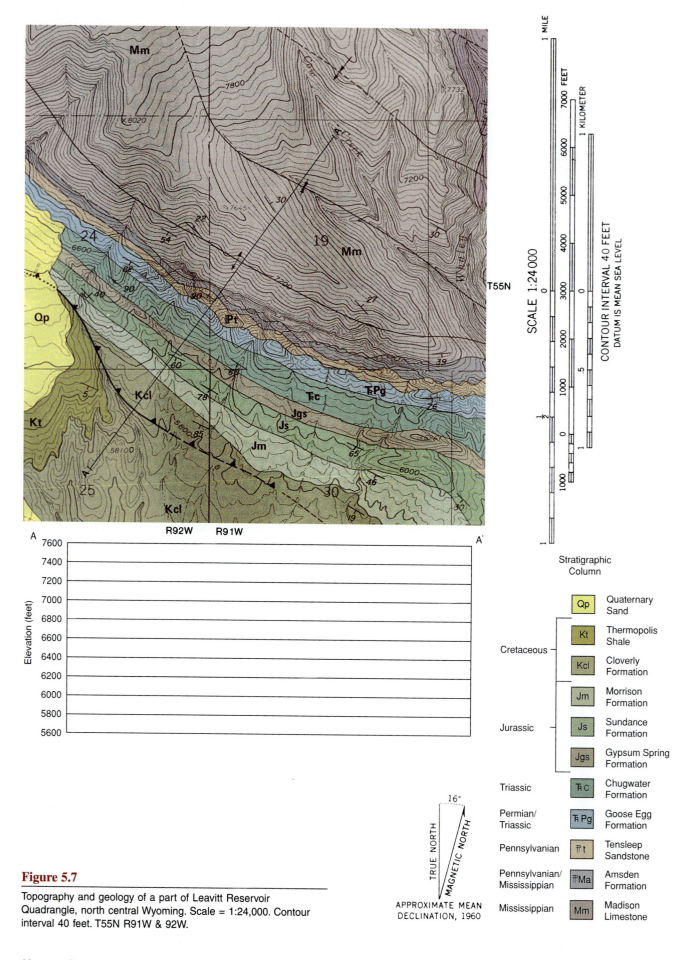

Figure 5.7

Topography and geology of a part of Leavitt Reservoir
Quadrangle, north central Wyoming. Scale = 1:24,000. Contour
interval 40 feet. T55N R91W & 92W.

SCALE 1:24 000

CONTOUR INTERVAL 40 FEET
DATUM IS MEAN SEA LEVEL

Stratigraphic
Column

	Qp	Quaternary Sand
Cretaceous	Kt	Thermopolis Shale
	Kcl	Cloverly Formation
	Jm	Morrison Formation
Jurassic	Js	Sundance Formation
	Jgs	Gypsum Spring Formation
Triassic	℞c	Chugwater Formation
Permian/ Triassic	℞Pg	Goose Egg Formation
Pennsylvanian	℞t	Tensleep Sandstone
Pennsylvanian/ Mississippian	℞Ma	Amsden Formation
Mississippian	Mm	Madison Limestone

16°

TRUE NORTH

MAGNETIC NORTH

APPROXIMATE MEAN
DECLINATION, 1960

9. Conglomerate was deposited on this surface, creating a second unconformity (labeled 2).

10. A fault cut through the entire sedimentary rock sequence, pushing the rocks on the north side up relative to the rocks on the south side.

11. All of the conglomerate unit and part of the shale unit were eroded to form the present flat surface. Note the mismatch of units on opposite sides of the fault.

Deciphering the geologic development of an area is important when searching for deposits of petroleum and natural gas. It also helps geologists understand why some sedimentary deposits of valuable metals, ores of uranium, lead, zinc, and copper, are located where we find them.

Problems

1. Construct a topographic and geologic cross-section along line A-A′ on the Leavitt Reservoir Quadrangle (Figure 5.7).

2. In studying an unconformity, how might you try to determine how long it took the surface to develop, that is, the amount of time the surface represents?

3. Locate an unconformity surface on the map in Figure 5.7.

4. Construct a geologic cross-section and describe the geologic development along line A-A[1].

5. Construct a geologic cross-section along line B-B[1]. Why does the cross-section along line A-A[1] look so different from the cross-section along line B-B[1]?

Further Reading/References

Bernknopf, R. L. and others, 1993. "Societal value of geologic maps." U.S. Geological Survey Circular 1111, 53 pp.

Maltman, A., 1990. *Geological Maps: An Introduction.* New York, Van Nostrand Reinhold, 184 pp.

McCall, J., and Marker, B. (eds.), 1989. *Earth Science Mapping for Planning, Development and Conservation.* Boston, Graham & Trotman, 268 pp.

6 EXERCISE

SOILS AND SOIL POLLUTION

Igneous and metamorphic rocks form at depths between five miles and thirty miles below ground level at very high temperatures and pressures. In addition, the chemical environment at these depths is very different from the environment at the surface. At depth, there is less oxygen and less water. Hence, when igneous and metamorphic rocks are pushed up to the surface, they react chemically with oxygen and water in our atmosphere to form new minerals that are adjusted to surface conditions. The result of these reactions is *soil.* Soils are composed of debris from crumbling rocks, decayed plants, and decayed animals (Figure 6.1).

Some soils are residual, remaining where they formed from underlying rocks. Other soils have developed on old stream or landslide deposits. But whether residual or not, all soils go through the same developmental stages and reflect the same variables: parent materials, climate, and time. Depending on parent materials and climate, one foot of soil might require between 100 and 100,000 years to form. Soil formation is faster in hot, humid climates than in cold, dry climates because chemical reactions are faster at higher temperatures and when water is abundant and continually renewed.

Soils are mixtures of sand, silt, and clay particles and are named based on the proportions of each size grade (Figure 6.2). The most agriculturally productive soils contain subequal mixtures of the three sizes. Although each size fraction has a function in soil productivity, soil stability, and soil pollution, the most important size fraction is the clay size, composed almost entirely of clay minerals and organic matter. Although organic matter normally forms only

Figure 6.1

Grass rooted in organic-rich dark-colored topsoil (A-horizon) on top of white limestone.

a few percent of a soil, its presence is critical for agricultural productivity and environmental considerations.

CLAY MINERALS

Clay minerals are very small, usually less than one micrometer in length and width (1/25,000 inch) and have thicknesses that are perhaps a hundred times smaller. A clay flake is shaped like a piece of paper. Because they are so small, they are very reactive chemically. They collect on their sheetlike surfaces an assortment of ions and serve as a storehouse of nutrients for plants, ions such as potassium, calcium, phosphorus (as phosphate ion), and nitrogen (as nitrate ion). The

Figure 6.2

Classification of soils according to grain size.

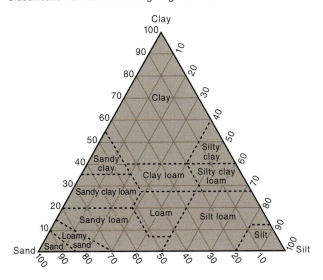

TABLE 6.1

Range of Concentration in Soils and Plants of Some Elements that Sometimes Occur as Environmental Contaminants

Element	Common Range in Concentration (ppm)	
	Soils	**Plants**
Arsenic	0.1–40	0.1–5
Boron	2–100	30–75
Cadmium	0.1–7	0.2–0.8
Copper	2–100	4–15
Lead	2–200	0.1–10
Manganese	100–4,000	15–100
Nickel	10–1,000	1
Zinc	10–300	15–200

Source: Data from N. C. Brady, The Nature and Properties of Soils, *8th edition, 1974, Macmillan Publishing Company, New York, NY.*

commercial fertilizer you scatter on your garden or lawn is composed largely of these elements and is formulated to supplement the inadequate supply of them in your soil.

Some of these nutrients are scavenged by the clay from the rainwater as it drips through the soil. The rainwater removes elements such as potassium and calcium from decaying feldspars and deposits them on clay surfaces. Other elements in soil water are obtained from decaying plant and animal tissues in the soil. All organic tissues are rich sources of phosphorus, and plants of the pea family (legumes) contain abundant nitrogen (as nitrate ion) captured from the air.

ORGANIC MATTER

In addition to a mixture of organic compounds dissolved in soil water, soils generally contain 1–6% organic matter, complex materials called *humus.* Humus is brown to black in color and is composed of organic compounds that are relatively resistant to decay. Dark-colored soils are rich in organic matter. Humus can hold a much larger amount of nutrient ions than the same amount of clay minerals. Plants rooted in soil that is deficient in organic matter will be poorly nourished.

In addition to its ability to trap nutrient ions for plants, organic matter holds abundant water and is sticky. The stickiness causes the soil to develop a "crumb structure," which greatly increases the ability of the air and water to penetrate the soil. Humus is essential to the development of a good agricultural soil.

SOILS AND POLLUTANTS

The surfaces of clay minerals and humus attract not only nutrient ions for plant nutrition but pollutants as well. Clays are negatively charged particles and therefore attract positively charged materials. Complex and harmful synthetic organic compounds such as PCBs and dioxins that fall on soil are adsorbed. So are *contaminant* heavy elements such as

arsenic, cadmium, and lead (Table 6.1), all of which are harmful to humans and many other living organisms. Soil can become seriously contaminated because of its ability to trap and hold many chemicals dissolved in the water that passes through them. On the other hand, contaminants held in the soil do not enter the water supply.

Some contaminants are held to clay and humus particles more tightly than others. Those held more tightly will displace those held less tightly in the competition for space on the host particle (clay or humus). The factors that determine the strength of attraction of dissolved substances for clay or humus are the size of the dissolved contaminant and its charge. Contaminants of larger size and higher charge will displace those that are smaller and have lower charge. For example, lead has an ionic radius of 1.26 angstroms (an angstrom is a very small unit used for ion sizes) and a charge of +2 (Table 6.2). Sodium ions have a radius of 0.98 angstroms and a charge of +1. Suppose a clay mineral is holding sodium ions on its surface and soil water containing lead ions passes by. The lead will replace the sodium on the clay and push it into the water to be carried away. This is a good result because the lead is removed from the water. Lead in drinking water can cause brain damage in humans.

But there is a limit to the amount of contaminant a soil can hold. If contamination continues for too long, the soil materials can become saturated and the contaminants will pass freely into our underground water supply (Table 6.3). The point at which a soil becomes saturated depends on the amounts and types of clay and humus it contains, and on the type of contaminant.

CLASSIFICATION OF SOILS

As they form from underlying rocks or sediments, soils develop a sequence of parallel layers called horizons

TABLE 6.2

Size and Charge of Some Important and/or Dangerous Inorganic Contaminants

Element	Radius	Charge
Selenium	0.35	+6
Arsenic	0.47	+5
Chromium	0.64	+3
Selenium	0.69	+4
Uranium	0.89	+4
*Sodium	0.98	+1
Cadmium	0.99	+2
*Calcium	1.04	+2
Mercury	1.12	+2
Lead	1.26	+2
*Potassium	1.33	+1

Note: Some ions occur in more than one size and charge. Those marked with an asterisk are not dangerous.

(Figure 6.3). They are labeled O, A, B, and C. The *O-horizon* is the layer of litter and decaying organic matter on the surface. It contains only very small amounts of mineral fragments but is a rich source of some kinds of dissolved nutrients. Below it lies the mineral-rich part of the soil, composed of more than 90% inorganic materials.

The *A-horizon* is also called the *topsoil* and is the zone in which the plants are rooted. As rainwater percolates through the decomposing mineral grains, it dissolves them, produces clays, and carries both the dissolved ions and the solid clays downward to be deposited lower in the soil profile. The A-horizon is referred to as the zone of leaching (removal). The zone below it in which the material from the topsoil may be deposited defines the *B-horizon* or *subsoil*. It is also called the zone of accumulation. Below the B-horizon, lies the *C-horizon,* a zone composed of broken and partially decomposed bedrock or parent sediment.

Soils in humid climates accumulate red iron oxide (hematite) in the B-horizon and are called *pedalfers*. Soils in dry climates accumulate white calcium carbonate (calcite) in the B-horizon and are called *pedocals*. The changeover from pedalfers in the humid eastern United States to pedocals in the dryer western United States occurs at an annual precipitation of about 30 inches. Both pedocals and pedalfers can be agriculturally productive soils, although pedocals usually need to be irrigated to supplement their deficient water supply. The isohyet of 30 inches runs north-south through mid-America, and most crop farms in this area need to be irrigated. The same is true of central California's San Joaquin Valley, where a large proportion of the nation's fruits and vegetables are grown.

TABLE 6.3

Typical Ranges of Heavy Metal Concentrations in Sewage Sludges, Fertilizers, Farmyard Manure, Lime, and Composts (ppm)

	Sewage Sludge	Phosphate Fertilizers	Nitrate Fertilizers	Farmyard Manure	Lime	Composted Refuse
Silver	<960	—	—	—	—	—
Arsenic	3–30	2–1,200	2.2–120	3–25	0.1–25	2–52
Boron	15–1,000	5–115	—	0.3–0.6	10	—
Cadmium	<1–3,410	0.1–170	0.05–8.5	0.1–0.8	0.04–0.1	0.01–100
Cobalt	1–260	1–12	5.4–12	0.3–24	0.4–3	—
Chromium	8–40,600	66–245	3.2–19	1.1–55	10–15	1.8–410
Copper	50–8,000	1–300	—	2–172	2–125	13–3,580
Mercury	0.1–55	0.01–1.2	0.3–2.9	0.01–0.36	0.05	0.09–21
Manganese	60–3,900	40–2,000	—	30–969	40–1,200	—
Molybdenum	1–40	0.1–60	1–7	0.05–3	0.1–15	—
Nickel	6–5,300	7–38	7–34	2.1–30	10–20	0.9–279
Lead	29–3,600	7–225	2–27	1.1–27	20–1,250	1.3–2,240
Antimony	3–44	<100	—	—	—	—
Selenium	1–10	0.5–25	—	2.4	0.08–0.1	—
Uranium	—	30–300	—	—	—	—
Vanadium	20–400	2–1,600	—	—	20	—
Zinc	91–49,000	50–1,450	1–42	15–566	10–450	82–5,894

Figure 6.3

A generalized soil profile. Individual horizons can vary in thickness. Some may be locally absent, or additional horizons or subhorizons may be identifiable.

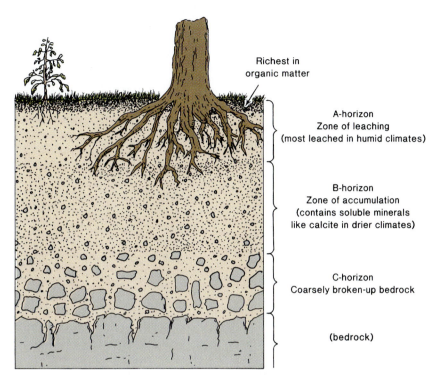

Richest in organic matter

A-horizon
Zone of leaching
(most leached in humid climates)

B-horizon
Zone of accumulation
(contains soluble minerals like calcite in drier climates)

C-horizon
Coarsely broken-up bedrock

(bedrock)

SOIL AND AGRICULTURE

The location of the world's "breadbaskets," the places where most of the world's food is grown, is determined mostly by climate. Farming without irrigation is most productive when the annual rainfall is 30–40 inches; with irrigation the range can be extended down to 15 inches. At lower annual rainfall there is insufficient soil development. The optimum average yearly temperature is 50–65° Fahrenheit. When the temperature is too low, chemical reactions are so slow that *photosynthesis* is retarded. If the temperature is too high, respiration is too fast and the plant grows more slowly. At temperatures above 110°, plant enzyme systems are destroyed.

Edible crops are those that not difficult to cultivate, have been found over thousands of years to be nourishing, and that we have grown accustomed to eating. In the Western world wheat is the favorite; in east Asia rice is the staple. Other grains grown extensively in the United States and elsewhere include corn, oats, barley, and rye. From an ecological viewpoint, it clearly is not desirable to deliberately grow one type of plant year after year over thousands of square miles, as is common in the United States. Monocultures such as this are more vulnerable to insect pests and also deplete the soil of essential nutrients to a greater extent than would otherwise occur.

World grain production appears to have stopped increasing after many decades of growth. There are several reasons for the change, including the growth of cities at the expense of farms, rerouting of irrigation water to the needs of city dwellers, and the continuing loss of topsoil to erosion and harmful salt accumulation *(salinization)*. Until 1950 all growth in world food output resulted from increases in the number of acres being farmed. By 1980 only 20% of the growth in food output resulted from farming more acres; 80% came from making the acreage more productive using new farming methods and chemicals. Since 1980 all increases in productivity have come from making the land work harder. The decreasing area devoted to farming, loss of topsoil to erosion, and pollution of the soil are slowly destroying America's agricultural base.

SOIL SURVEYS AND LAND-USE PLANNING

What is the best use for soil in a particular area? How much can be farmed? Does the soil contain enough organic matter for efficient farming? Where should homes be built? Does the soil contain the type of clay that expands to several times its volume, resulting in cracked foundations? Is the drainage of water through the soil too poor to permit road construction? Such questions can be answered by the construction of soil maps. Some maps show soil types, other soil thicknesses, and others may show the percentage of swelling clays. The U.S. Geological Survey and many state and local governments have published soil maps. The number of such maps is increasing as the public becomes more aware of soil, one of our most important but deteriorating natural resources.

Problems

1. Do you think a soil will develop faster on new granitic bedrock or on an ancient stream sand that has the same mineral composition? What are the differences between the two that led to your answer?

2. The best soil is a loam. Why does the suitability of the soil for agriculture decrease if the percentage of sand increases? Or if the percentage of clay increases?

3. What is humus? Why is it important from an agricultural point of view?

4. Construction workers have added large amounts of slag (glass from an ore processing plant) containing 2,000 ppm of lead to a residential area to level the uneven ground surface. What factors should be considered to evaluate the likelihood that the potentially harmful lead atoms will be freed from the slag and absorbed onto soil material?

5. Suppose the slag contained uranium rather than lead. In what ways would this change your approach to answering the previous question?

6. Why are only positively charged ions absorbed in significant amounts by soils?

7. From Table 6.1 it is clear that plants generally contain lesser amounts of the eight elements listed than does the soil in which the plants grow. Based on this observation, what might you infer about the chemistry of plant growth?

8. Arsenic, cadmium, and lead are listed in both Table 6.1 and 6.2. How might you explain the differences in abundance of these three elements in soils.

9. America's major mountainous areas are the Appalachians in the east and the Rockies, Basin and Range area, and Sierra Nevada in the west. Find the map in your textbook that shows the isohyets in the United States. Based on these data, where would you predict America's major agricultural areas would be located. Why? Where might irrigation be needed?

10. Where in the western United States would you expect pedalfers to form? Are there areas in the eastern half of the country where you would expect to find pedocals?

11. Shown below are three graphs illustrating the effect that an increasing thickness of topsoil, increasing amount of organic carbon, and increasing amount of erosion have on crop yield in America's agricultural areas. Explain these trends.

12. In a famous experiment, J. B. von Helmont (1577–1644) investigated the source of the materials that plants are composed of.

> That all vegetable [matter] immediately and materially arises from the element of water alone I learned from this experiment. I took an earthenware pot, placed in it 200 lb of earth dried in an oven, soaked this with water, and planted in it a willow shoot weighing five lb. After five years had passed, the tree grown therefrom weighed 169 lb and about 3 oz. But the earthenware pot was constantly wet only with rain or (when necessary) distilled water; and it was ample [in size] and imbedded in the ground; and, to prevent dust flying around from mixing with the earth, the rim of the pot was kept covered with an iron plate coated with tin and pierced with many holes. I did not compute the weight of the deciduous leaves of the four autumns. Finally, I again dried the earth of the pot, and it was found to be the same 200 lb minus about 2 oz. Therefore, 164 lb of wood, bark, and root had arisen from the water alone.

This was an excellent experiment for the 17th century— well-planned, carefully done, and accurately described. Do you agree with von Helmont's conclusion? Why or why not? What factors was he unaware of in the early 17th century that might have led him to design the experiment differently?

The effect of A-horizon thickness, amount of organic carbon in a soil, and amount of erosion on crop yield.

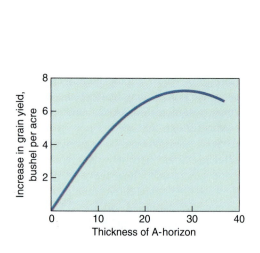

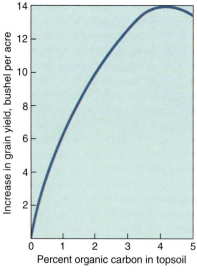

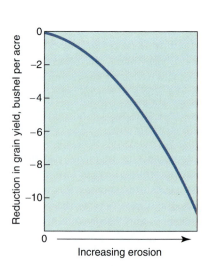

Further Reading/References

Brady, N. C., and Weil, R. R., 1996. *The Nature and Properties of Soils,* 11th ed. New York, Macmillan, 621 pp.

Glanz, James, 1995. *Saving Our Soil.* Boulder, Colorado, Johnson Books, 182 pp.

Hillel, Daniel, 1991. *Out of the Earth.* Berkeley, California, University of California Press, 321 pp.

Pierzynski, Gary M., Sims, J. Thomas, and Vance, George F., 1994. *Soils and Environmental Quality.* Boca Raton, Florida, Lewis Publishers, 313 pp.

Troeh, Frederick R., and Thompson, Louis M., 1993. *Soils and Soil Fertility,* 5th ed. New York, Oxford University Press, 462 pp.

7 EXERCISE

SWELLING SOILS

When minerals in igneous and metamorphic rocks alter to other substances at the earth's surface, most of the materials produced are harmless to humans and their civilization. Many of them are not only harmless but are beneficial. Examples include plant nutrients such as potassium, phosphorus, and iron. There are, however, a few substances produced by weathering that cause construction or pollution problems. One of these is montmorillonite clay.

SWELLING CLAY

Montmorillonite is one of three common types of clay minerals. It is composed mostly of silicon and aluminum, as are all clays, but is distinctive because it contains sodium and calcium as essential elements. It forms from the chemical alteration of rocks rich in these elements, such as basalt and its fragmental volcanic equivalent, dark-colored volcanic tuff. Montmorillonite differs from other clays because it has an exceptionally strong attraction for water molecules. Clays have a sheetlike crystal structure, and the water molecules force their way between the sheets and cause an enormous expansion of the clay flakes. The tiny clay particles can expand to 20 times their dry volume. The water that causes the expansion can come from rain and snow, watering the lawn and shrubbery, leaking water pipes, or crop irrigation.

If montmorillonite is abundant in a soil, the soil swells or shrinks, depending on the availability of water, with devastating results to buildings (Figures 7.1, 7.2). A dry montmorillonitic soil reveals that it has shrunk by the presence of a soil surface described as popcorn topography (Figure

7.3). These hazardous soils are called *vertisols* by soil scientists and expansive or *swelling soils* by engineers. Pure montmorillonite can expand up to 2000%, and the pressure generated by the expansion can be 10 tons per square foot. However, few soils are pure montmorillonite and many soils do not contain the mineral at all. In most montmorillonitic soils the amount of expansion is only 20–50%. But an expansion of only 3% is considered by construction engineers to be hazardous to buildings. A pressure of only one ton per square foot is greater than the load exerted by small buildings, which can be uplifted and rotated by the expanding soil. Foundations and walls crack and separate.

Damage caused by expansive soils is widespread wherever the soil contains a significant amount of basaltic volcanic materials. The distribution of swelling soils in the United States is shown in Figure 7.4. Most severely affected are the northern midcontinent and the Gulf Coastal area of east Texas and southwestern Louisiana, but about one-third of the conterminous United States contains soils rich in montmorillonite clay. Several hundred thousand new homes are built each year on expansive soils. Damage caused by swelling soils now totals about $8 billion annually.

Problems

1. Using Figure 7.4, examine a geologic map of the United States and locate the geologic formations where swelling clays are a serious problem. Are they predominantly sandstones, shales, or limestones? Is this what you would expect? Why?

Figure 7.1

Examples of soil problems. (a) Slab poured in dry season, soil expansion at periphery during wet season; (b) slab poured in wet season, soil shrinkage at periphery in dry season; (c) building supported by cut and fill subject to differential expansion and contraction. *From Gary B. Griggs and John A. Gilchrist,* Geologic Hazards, Resources, and Environmental Planning, *2d ed. Copyright © 1983 Wadsworth Publishing Company, Belmont, CA. Reprinted by permission.*

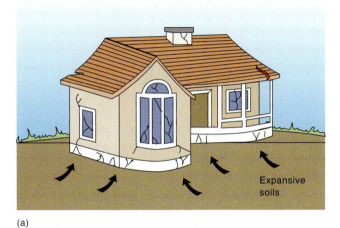

(a)

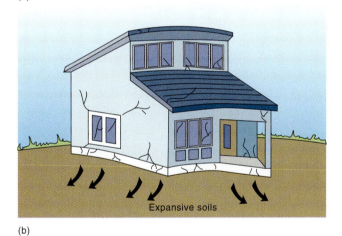

(b)

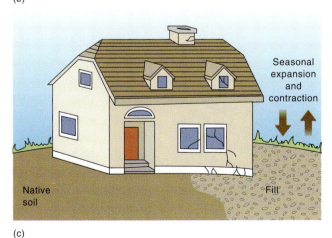

(c)

2. Based on the behavior of montmorillonite clay in water, what might you infer about the bonding between adjacent clay sheets in this mineral?

Figure 7.2

Building wall damaged by stair-step corner fracturing. Lakewood, Colorado, October 1976. The center section of the wall at right has been uplifted relative to the corner at left. Damage of this sort is frequently caused by expansive clay.

3. Why do you think basaltic parent rocks are more likely to give rise to swelling soils than are other rock types? Name another type of igneous rock that is likely to yield swelling soils. What metamorphic rocks will weather to produce montmorillonite? (Hint: use Table 2.1 and Figure 3.12.)

4. Could a sandstone weather to produce a montmorillonitic soil? If you believe it could, explain the requirements for it to occur.

5. Would you expect to find swelling soils more common among pedalfers or pedocals? Why?

6. A rock composed almost entirely of montmorillonite clay is called bentonite. It is used in oil well drilling as a lubricant to cool and preserve the drill pipe and drill bit during exploration for petroleum. Why do you think clay is used at all rather than simply circulating only water? Why do you think montmorillonite clay is used rather than one of the other two types of clay?

Figure 7.3

"Popcorn-like" texture of soils at the ground surface characterized by bulges and cracks is a characteristic of soils with swelling clays. This Wyoming soil is full of bentonite, a deposit of swelling clays mined commercially for many uses—from cosmetics to oil well drilling fluids.

Figure 7.4

Distribution of swelling soils within the conterminous United States. Areas in red are most abundant in swelling soils, followed by areas in blue. Uncolored areas are not devoid of swelling soils. There are many small local occurrences that are not shown. *From E. B. Nuhfer, et.al.,* The Citizen's Guide to Geologic Hazards. *Copyright © 1993 American Geological Institute, Alexandria VA. Reprinted by permission.*

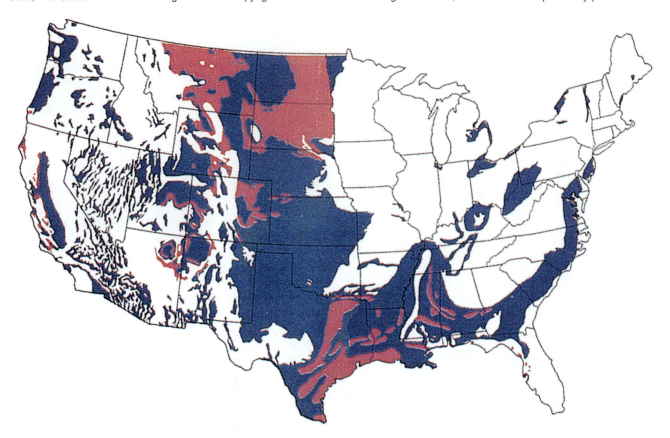

7. Examine the state or local geologic map provided by your instructor. Where do you think there could be a problem with swelling soils? Suppose you were considering building your dream house in a suspect area. Explain how you might determine whether a problem does in fact exist. How would you locate someone to do a site investigation? How much would it cost? Phone a suitable company and ask. How would you ask this company to sample the soils? How might you decide the percentage of swelling clay you would be willing to tolerate in the soil under your house?

Further Reading/References

Holtz, W. G., and Hart, S. S., 1975. "Home construction on shrinking and swelling soils." *Colorado Geological Survey Special Publication 11,* 18 pp.

Jochim, C. L., 1981. "Home landscaping and home maintenance on swelling soil." *Colorado Geological Survey Special Publication 14,* 31 pp.

Jones, D. E., and Holtz, W. G., 1973. "Expansive soils, the hidden disaster." *Civil Engineering,* v. 8, p. 49–51.

Tourtelot, H. A., 1974. "Geologic origin and distribution of swelling clays." *Bulletin of the Association of Engineering Geologists,* v. 11, p. 259–75.

8

SLOPE STABILITY AND LANDSLIDES

Landslides or *mass wasting* always occurs on slopes and consists of downslope movement of soil and bedrock. Damage resulting from these movements totals $1.5 billion annually, and loss of life can be catastrophic. Thousands of people have been killed by individual landslides in many countries. Fortunately, mass movements of rock, loose sediment, and soil are one of the more controllable natural disasters, given modern technical knowledge and public awareness.

Mass movements are distinguished from streamflow by the relative amounts of water and solid material. In a stream the amount of solid material in motion is usually only one percent or so of the sediment-water volume. In a mass movement or landslide, the percentage of rock and sediment is greater than 50% and in avalanches and rockslides is 100%.

CAUSES OF MASS MOVEMENTS

The strength or resistance to movement of a mass of rock is controlled by the extent and spacing of discontinuities within the rock mass. Examples of naturally occurring discontinuities in rocks include (1) depositional features such as bedding and fissility; (2) erosional features such as unconformities and scours; (3) metamorphic features such as slaty cleavage and schistosity; (4) *tectonic* features such as joints and faults; and (5) mineralogic features such as the presence of clay minerals. All these features are structural weaknesses and lead to mechanical or chemical loss of cohesion between rock surfaces. Because such discontinuities exist everywhere, unstable slopes and mass movements can occur in all terrains and climates.

All that is needed for a mass movement to occur is a triggering mechanism strong enough to overcome the rock-rock or rock-soil friction that maintains the existing slope. Naturally occurring triggers include earthquakes, volcanic eruptions, rainfall, snowmelt, and undercutting by stream erosion or sea waves. Landslides can be grouped into three types based on the amount of water involved and the cohesiveness of the movement (Figure 8.1). The fastest moving and most dangerous to life are noncohesive types of movements: rockfalls or avalanches. These are downslope movements of dry rock at speeds of up to several hundred miles per hour. Boulders and large blocks of rock toppling from a cliff face and falling freely through the air are rockfalls. Only slightly less dangerous are debris slides, rapid downslope movements of rock and sediment along the ground surface. When loose debris on a slope is destabilized by rainfall or an earthquake, slides result. Cohesive movements called slumps, creeps, or earthflows consist of a downslope movement of cohesive soil and sediment.

Resistance to slope failure results from two factors. The first is internal friction caused by grain-to-grain contacts between irregularly shaped, coarse granular particles. Any granular material has the ability to stand at some angle or maintain some slope because of grain-to-grain friction. The second factor is cohesion caused by materials that hold particles together in a solid, impermeable mass, such as occurs in clay. Common binding forces are electrostatic attraction or chemical bonds. Shales exemplify this type of cohesion.

Studies of potential slope failure in coarse granular rocks usually focus on joint spacing. Along joint surfaces

Figure 8.1

Examples of landslides by type of movement. *Source: W. W. Hays (editor), 1981, Facing Geologic and Hydrologic Hazards: Earth Science Considerations. U.S. Geological Survey* Professional Paper 1240-B, *p. 61.*

Noncohesive Movements

Fall (rock fall)

Fall – Mass travels most of the distance in free fall.

Slide (rock slide)

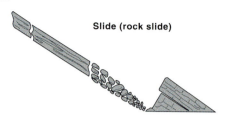

Translational Slide – Movement is predominantly along planar or gently undulatory surfaces, frequently controlled by surfaces of weakness, such as faults, joints, or bedding planes.

Slide – Movement of material by shear displacement along one or more surfaces or within a relatively narrow zone.

Cohesive Movements

Slide (rock slump)

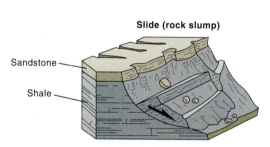

Sandstone

Shale

Rotational Slide – Movement involves turning about a point (surface of rupture is concave upward).

Flow (earth flow)

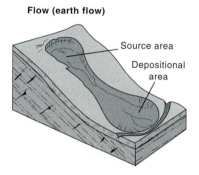

Source area

Depositional area

Flow – Movement of mass such that the form taken by moving material resembles that of viscous fluids.

the two rock halves make contact only at surface irregularities. These contacts are not like those of an interlocking picture puzzle, so the only force maintaining rock stability and cohesion is grain-against-grain friction. This frictional resistance to slab movement can be decreased by shaking (earthquakes), by water percolation through discontinuities, or by the 9% expansion of water as it freezes. Percolating water decreases grain-to-grain friction not only by increasing pore pressure but also by causing chemical alteration (weathering) of these surfaces, softening them through the formation of clay minerals. Clay minerals decrease cohesion because they are soft and easily gouged, are sheetlike in character, and absorb water. As the proportion of clay minerals in an aggregate of grains increases, the type of slope failure changes from noncohesive to cohesive.

Most soil failures (earthflows, slumps) are cohesive because most soils contain high percentages of clay. In general, the greater the quantity of clay minerals in a soil, the greater its potential for shrinkage and swelling, the higher its plasticity, and the greater the likelihood that slope failure will be cohesive. Consequently, studies of soil strength

and resistance to downslope movement usually focus on clay content. The three types of clay, kaolinite, illite, and montmorillonite, absorb different amounts of water. Montmorillonite is by far the greatest water-absorber so that clay aggregates or shales that contain even small percentages of montmorillonite swell, lose cohesion, and slide easily even on the gentlest of slopes.

Montmorillonite has a high liquid limit, meaning that plastic behavior begins at low clay contents. It also has a high plasticity limit, meaning that the clay-water mixture maintains plasticity at very high clay/water ratios, a result of large amounts of bound water in montmorillonite compared to the other clay types, illite and kaolinite.

Ground failure can also result from processes unrelated to discontinuities in solid rock or to the amount of montmorillonite in clay-rich materials. Two of these processes are *liquefaction* and *spreading failures.* Liquefaction occurs when granular materials change in behavior from solid to liquid in response to increased pore pressure. This occurs when ground-shaking reorients the unconsolidated sediment grains into a more compact arrangement. When the groundwater table is near the surface during this reorientation, the shaking reduces the grain-to-grain contacts, and the load is temporarily transferred to the pore water. These changes increase pore pressure and decrease grain-to-grain friction, and the deposit then behaves as a liquid.

Liquefaction beneath a layer of hard rock can cause the overlying, firmer rock or soil to break into units and spread apart. Spectacular ground collapses occur when a fine-grained sediment changes from a fairly hard, strong, brittle solid to a liquid of negligible strength. Materials exhibiting this unusual property are termed *quick clay* or *quicksand,* although some of these sensitive sediments contain little or no clay. The clay flakes are arranged in a fluid-filled, house-of-cards structure; when the aggregate is disturbed the structure collapses, and the material behaves like a fluid.

RECOGNIZING AND PREVENTING SLOPE FAILURES

Various studies have shown that the most damaging slope failures are closely related to human activities; regulating land use *before* these activities take place can substantially reduce loss. The old adage "an ounce of prevention is worth a pound of cure" applies well to slope failures.

Recognizing Areas Susceptible to Slope Failure

No area is immune to slope failure, but some areas are more likely to experience catastrophic failures than others. What should you look for?

1. Steep slopes. Gravity is the force that causes downslope movement, so avoid steep slopes. The gentler the better.

2. Bedrock. Rocks that are relatively incoherent, such as poorly cemented sandstones, are dangerous. Rocks that have planes of weakness are dangerous. These planes include faults, bedding, foliation, fissility, and jointing. Slopes that are parallel to planes of layering or jointing are particularly unstable.

3. Areas that lack vegetation. Plant roots bind soil and stabilize it. The more trees, the better. Bare areas are much more likely to slide downhill catastrophically than vegetated areas.

4. Areas of very heavy rainfall. Precipitation increases the pore pressure and separates the grains. It also makes soil soggy so that it will flow plastically.

5. Slope undercutting. Landslides are particularly common along stream banks, highway road cuts, and seacoasts during storms, all places where the natural support for a slope has been removed.

6. Areas that have had earthquakes relatively recently. Shaking will destabilize slopes by decreasing the friction provided by grain-to-grain contacts. The result can be wholesale slope failure when a steep gradient is present, or liquefaction when the ground is flat and the substrate is unconsolidated sediment.

Because scientists and engineers understand the factors that make slopes unstable (Figure 8.2), prevention is an attainable goal in many regions. Dangerous areas can be (1) avoided if possible; (2) stabilized by structures that in-

Figure 8.2

Cross-section showing homeowners with no hope. Their house is built on a combination of unstable fill and shale whose layering is parallel to the ground slope. In addition, the shale and sandstone behind the house are unstable, and the sandstone feeds water into the planes of fissility in the overlying shale; sinkholes are developing in the limestone below the house and will soon undermine the house; and the toe of the fill supporting the house is being undermined by water leaking from the sandstone.

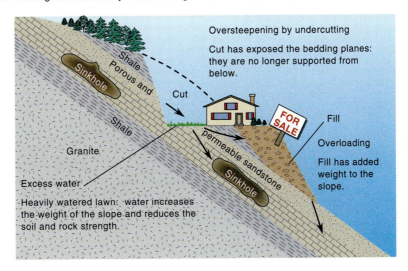

Figure 8.3

Chain-link mesh keeps loose rocks from rolling into the highway.

clude retaining walls, drainage ditches, subsurface pipes, and wire mesh coverings (Figure 8.3); or (3) stabilized by grading slopes or planting trees, whose roots improve soil cohesion.

Approximately 25 years ago the U.S. Geological Survey began engineering-geology mapping projects in the San Francisco Bay area. This region was chosen because it is prone to both earthquakes and landslides and is heavily populated. The results (for example, Figure 8.4) are helpful to both engineering geologists concerned with construction and land-planners concerned with a wider range of possible land uses. Since 1970 many state governments have commissioned maps for various environmental purposes such as landslide control, groundwater safety, and soil erosion and pollution. The number and variety of such maps is increasing rapidly as public concern about environmental problems increases.

Problems

1. Hollidaysburg is a town of 6,000 people in southwestern Pennsylvania near the confluence of two rivers and surrounded by mountains with up to 450 feet of local relief. Because of this location, considerable potential for mass movements of surficial materials exists within five miles of the town. The town government has established a committee to determine the extent of danger from natural disasters. As a member of this committee and an educated person with some training in environmental geology, you have obtained a geologic and topographic map of the area (Figure 8.5). Based on this map, you are asked to answer the following questions:

   a. What is the range in geologic age of the rocks exposed in the map area?

   b. What type of geologic contact is present between Qal and Dha?

   c. What type of sedimentary rock—sandstone, shale, or limestone—is most abundant in the area? Least abundant? What do your answers suggest about the frequency of mass movements in the map area?

   d. Identify the formation that should be the most resistant to mass movement. What are the characteristics that caused you to choose it?

   e. In what way will the relative proportions of the three rock types affect the useful life of the reservoir located 1.5 miles SSE of the center of Hollidaysburg?

   f. What is the ground-slope angle of the hill immediately south of the reservoir? What is the dominant type of rock exposed on the hill?

   g. The southwestern part of the city is built on Qal. What environmental problems do you think this might lead to?

   h. At the northwest edge of the town of Newry is a steep hill underlain by the formation labeled Db. Do you think the town faces any danger from this hill? Explain.

   i. Most of Hollidaysburg is built on Sc and Smk. Might this be a possible problem for the inhabitants located closest to Beaverdam Branch? Explain.

   j. Is there any danger to the railroad track from the hill just east of Kladder Station, located 0.8 inches north of the southern end of the map? Explain.

Figure 8.4

(a) Slope-failure map of the Congress Springs area, Santa Clara County, California. (b) A derivative map showing potential ground movement and recommended land-use policies. The original maps are on a topographic map base at a scale of 1 inch to 250 feet (1:3,000). *Source: U.S. Geological Survey, 1982, pp. 24, 32.*

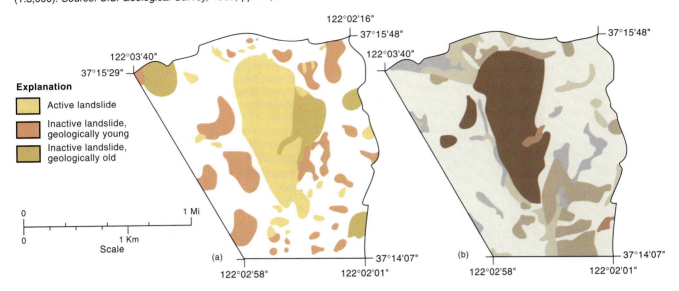

Relative Stability	Map Area	Geologic Conditions	Recommended Land Use		
			Houses	Roads	
				Public	Private
Most Stable		Flat or gentle slopes; subject to local shallow sliding, soil creep and settlement	Yes	Yes	Yes
↑		Gentle to moderately steep slopes in older stabilized landslide debris; subject to settlement, soil creep, and shallow and deep landsliding	Yes*	Yes*	Yes*
		Steep to very steep slopes; subject to mass-wasting by soil creep, slumping and rock fall	Yes*	Yes*	Yes*
		Gentle to very steep slopes in unstable material subject to sliding, slumping, and soil creep	No*	No*	No*
↓		Moving, shallow (<10 ft) landslide	No*	No*	No*
Least Stable		Moving, deep landslide, subject to rapid failure	No	No	No

Yes* - The land use would normally be permitted, provided the geologic data and/or engineering solutions are favorable. However, in some instances the use would be inappropriate.

No* - The land use would normally not be permitted. However, under some circumstances geologic data and/or engineering solutions would permit the use.

k. Can you suggest one or two reasons why the stream valley is widest about one mile ESE of Newry?

l. What is the general relationship between the dip of the sedimentary rocks, the types of rock, and the propensity for mass wasting?

2. Describe the various ways in which water is involved in slope failures.

3. Discuss the relative effects of rock characteristics and climate on the occurrence of slope failures.

4. List the factors that affect whether an earth movement will be slow, such as an earthflow, or fast, such as an avalanche.

5. How might you distinguish between mass-movement deposits and stream or glacial deposits?

6. Volcanoes are commonly sites of massive earth movements. List reasons why the presence of a volcano or group of volcanoes increases the risk of such occurrences.

7. The state highway department is considering constructing a new highway along the base of a hill composed of limestone. What kinds of environmental problems might the construction generate? How should they be investigated and remedied?

Further Reading/References

Brabb, E. E., 1991. "The world landslide problem." *Episodes,* v. 14, p. 52–61.

Costa, J. E., and Wieczorek, G. F. (eds.), 1987. *Debris Flows/Avalanches: Process, Recognition, and Mitigation.* Reviews in *Engineering Geology,* v. VII,

Figure 8.5

Geologic map of the area surrounding Hollidaysburg, Pennsylvania.

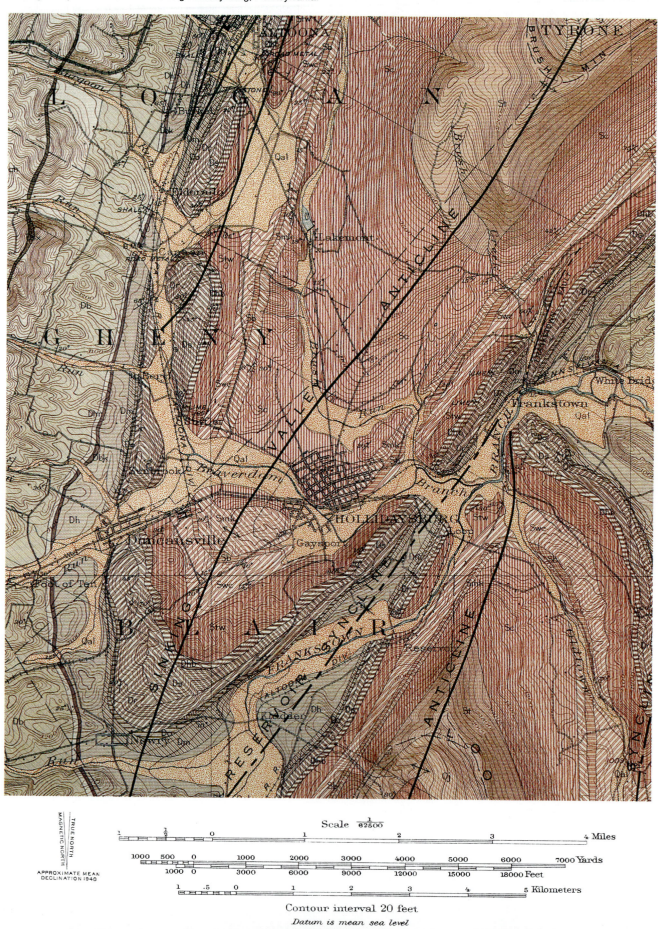

Scale $\frac{1}{62500}$

Contour interval 20 feet

Datum is mean sea level

Edition of December 1945

Figure 8.5—*Continued*

EXPLANATION
SEDIMENTARY ROCKS

Qal

Alluvium
(silt, sand, and gravel constituting the flood-plains of present streams)

Recent — QUATERNARY

Dha

Hampshire formation
(predominantly red, lumpy shale or mud rock and red sandstone; some gray and green shale and sandstone)

Dch **Dsx**

Chemung formation
(chiefly green, gray, and chocolate-colored shale and thin beds of argillaceous fine-grained sandstone; fossiliferous throughout; includes Saxton conglomerate member. Dsx: upper part largely chocolate-colored)

Db

Brallier shale
(micaceous, siliceous slaty green shale with some thin beds of fine-grained sandstone; sparsely fossiliferous throughout, mainly pelecypods of Gardeau type)

Dhr **Dbk**

Harrell shale
(soft gray shale in upper part; Burket black shale member, Dbk, in lower part; highly fossiliferous, small pelecypods and cephalopods of the Naples fauna)

Upper Devonian — *Portage group*

Dh

Hamilton formation
(principally olive-green shale with even-layered, blocky-jointed sandstone and thin limestone at top; ridge-making sandstone at two horizons; sparingly fossiliferous; locally a foot or two of limestone at top with Tully fauna)

Dm

Marcellus shale
(black fissile clay shale, grading upward into olive-green shale)

Do

Onondaga formation
(gray shale, probably calcareous, and thin argillaceous limestone)

Middle Devonian

Dr

Ridgeley sandstone
(thick-bedded calcareous sandstone weathering to coarse friable sandstone; locally a fine conglomerate at top with quartz pebbles; highly fossiliferous)

Ds

Shriver limestone
(thin-bedded siliceous limestone, weathering to fine-grained sandstone; black calcareous shale at bottom; sparingly fossiliferous)

Dhb

Helderberg limestone
(lower part is thick-bedded gray limestone with thin gray chert at top, chiefly Keyser limestone member; overlying Coeymans and New Scotland limestone members thin and locally absent; contains valuable quarry rock, called "calico rock"; fossiliferous throughout)

Lower Devonian — *Oriskany group* — DEVONIAN

Stw

Tonoloway limestone
(thin-bedded finely laminated, dark limestone; sparingly fossiliferous, chiefly Leperditia)

Swc

Wills Creek shale
(chiefly gray, calcareous shale and some greenish limestone; fossils scarce)

Sb

Bloomsburg redbeds
(lumpy red shale and thick-bedded ridge-making red sandstone)

Smk

McKenzie formation
(blue thin-bedded fossiliferous limestone and soft gray and green shale; thin red shale east of Tussey Mountain and a little red shale west of Lock Mountain)

Cayuga group

Sc **Sk** **Scs**

Clinton formation
(mainly green and blue shale, weathering purplish, and thin fine-grained green sandstone in middle; Keefer sandstone member, Sk, near top; shale with thin limestone layers above Keefer sandstone member represent Rochester shale; Marklesburg iron-ore bed just beneath Keefer sandstone member; Frankstown iron-ore bed in lower half; hard quartzitic sandstone, red sandstone, and Levant Black iron ore, Scs at base; generally fossiliferous)

Niagara group — SILURIAN

St

Tuscarora quartzite
(hard white quartzite and sandstone, largely thick-bedded; quartzite extensively quarried for ganister; contains scolithus worm tubes and Arthrophycus at top)

Oj

Juniata formation
(chiefly red and some green fine-grained cross-bedded sandstone and red lumpy mud rock; nonfossiliferous)

Oo

Oswego sandstone
(gray fine-grained thick-bedded cross-laminated sandstone; contains a few small quartz pebbles in lower part; nonfossiliferous)

Upper Ordovician — ORDOVICIAN

Boulder, Colorado, Geological Society of America, 239 pp.

Matti, J. C., and Carson, S. E., 1991. *Liquefaction Susceptibility in the San Bernardino Valley and Vicinity, Southern California: A Regional Evaluation.* U.S. Geological Survey Bulletin 1898, 53 pp.

Mears, A. I., 1979. *Colorado Snow-Avalanche Area Studies and Guidelines for Avalanche-Hazard Planning.* Colorado Geological Survey Special Publication 7, 124 pp.

Nilsen, T. H., et al., 1979. *Relative Slope Stability and Land-use Planning in the San Francisco Bay Region,* *California.* U.S. Geological Survey Professional Paper 944, 96 pp.

Schuster, R. L., Varnes, D. J., and Fleming, R. W., 1981. "Hazards from ground failures; Landslides." In *Facing Geologic and Hydrologic Hazards, Earth-Science Considerations,* pp. 54–65. U.S. Geological Survey Professional Paper 1240-B.

Wold, R. L., Jr., and Jochim, C. L., 1989. *Landslide Loss Reduction: A Guide for State and Local Government Planning.* Colorado Geological Survey Special Publication 33, 50 pp.

E X E R C I S E 9

RIVER PROCESSES

Water is the most essential substance on earth. Without water to drink, all animals die in a few days. Plants wilt and decay. The amount of water on earth is immense, but nearly all of it is in the ocean and too salty to be used by those of us who live on land (Figure 9.1). If the earth's water supply totaled 50 gallons, the amount of immediately usable water would be about a teaspoonful.

Nevertheless, this metaphorical teaspoonful of water is able to carve the land surface into a bewildering variety of shapes and cause floods that ravage the landscape it has carved. Both normal streamflow and its excess—flooding—are of great importance to human societies. In this exercise we consider normal streamflow. In the next exercise we will examine flooding.

Streams are one of the most important features on the earth's surface. Stream courses that are dry throughout most of the year, bearing water only during and immediately after a rain, are called *ephemeral.* Stream channels that carry water during one part of the year, are dry during the other, and are fed by underground water are called *intermittent.* Streams that carry water continuously and are fed both by overland flow and from below are called *perennial.* The more humid the climate, the higher the proportion of perennial streams.

Total runoff on the ground surface depends on rainfall, evaporation, transpiration by plants, and infiltration into the soil. In general, the higher the amount of rainfall, the higher the runoff. For many communities, the runoff that supplies lakes and rivers controls both the quantity and the quality of the local water supply. If local rivers or lakes have a high enough capacity to supply the needed volume of water all

year, a community can draw from this source continuously. If water requirements exceed minimum streamflow, the community can store water until needed, although ponding increases the surface area of the water and therefore the amount of evaporation, reducing the ultimate yield. If its surface water supplies are inadequate, the community must tap underground sources. Water supply places the ultimate limitation on the number of people who can live in an area.

The catchment area of a stream is called its *watershed* or *drainage basin* and includes the entire area the stream serves (Figure 9.2). The famous Continental Divide in the western United States is an imaginary line that separates water channels that drain into the Pacific Ocean from those that drain into the Atlantic. Successively smaller drainage basins on the Atlantic Ocean side include the Gulf of Mexico basin, the Mississippi River basin, the Arkansas River basin, the Ohio River basin, and so on, down to the small drainage basin of the creek that flows by your house. Both the size and the shape of a drainage basin are important in determining runoff. Basin size, of course, has a strong effect on total runoff; the larger the catchment area, the greater the runoff. Basin shape is also important, however, because it influences the temporal distribution of runoff. In an elongate basin, flow in tributary channels reaches the main stream at different times, distributing the runoff over a long time span. In a more equant drainage basin, tributaries feed into the main stream at about the same time, resulting in a sudden high peak flow.

Topography also affects runoff, which decreases on gentler slopes. Gentle slopes allow more time for infiltration and for at least temporary water storage. They also tend to be

Figure 9.1

The hydrologic cycle. More than 99% of all earth's water is either in the oceans or frozen in glaciers. Although the percentages in each site are constant, the water molecules are continually cycled between the ocean, atmosphere, and land surface.

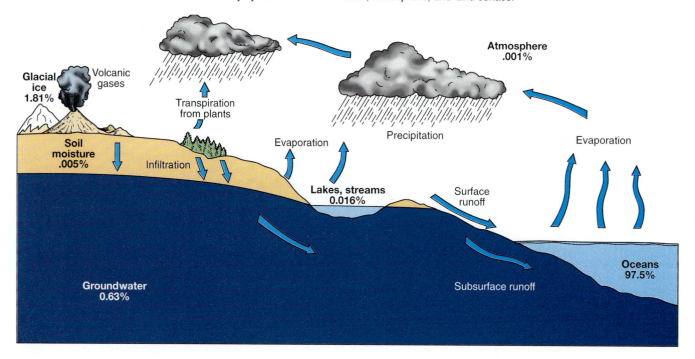

Figure 9.2

Adjacent drainage basins and the divide that separates them.

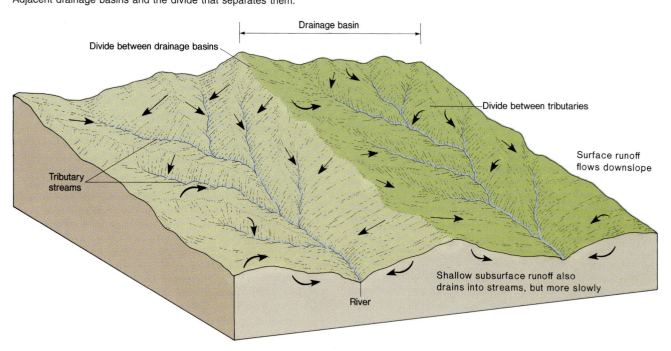

more densely vegetated than extremely steep slopes, and vegetation decreases runoff because plant roots hold water. Soil and surface-sediment character can be an important factor as well; loose, permeable sediment permits easy infiltration of water and thus decreases runoff.

HYDROGRAPHS

A *hydrograph* shows how streamflow varies with time and, therefore, reflects rainfall duration and intensity as well as the characteristics of the drainage basin that influence runoff. Hydrographs are plots of water discharge versus time

Figure 9.3

Long-period hydrograph for one point on Horse Creek near Sugar City, Colorado, showing the stream discharge over a period of one year. *Source: U.S. Geological Survey Open-File Report 79-681.*

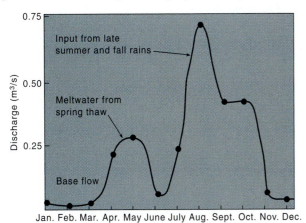

(Figure 9.3) that permit environmental geologists to determine the total flow, the base flow (upward flow from the groundwater into the stream), and periods of high (flood) and low flows. Long-period hydrographs (with graph axes calibrated in months or years) are used in designing irrigation projects and dam construction and for forecasting floods. Short-period hydrographs (with axes calibrated in hours or days) are used to show peak discharges during floods (Figure 9.4).

EFFECTS OF WATER MOVEMENT

The movement of channelized water has three consequences important to environmental geologists: erosion, sediment transport, and sediment deposition. Moving water has the ability to transport sediment; the amount transported depends on

Figure 9.4

Short-period flood hydrographs for two different points along Calaveras Creek near Elmendorf, Texas. The flood has been caused by heavy rainfall in the drainage basin. In the upstream part of the basin, flooding quickly follows the rain. The larger stream lower in the drainage basin responds more sluggishly to the input. *Source: U.S. Geological Survey Water Resources Division.*

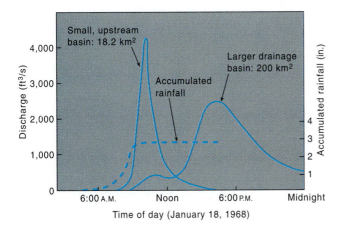

Time of day (January 18, 1968)

the amount of water, its velocity and gradient, and other factors. The moving sediment causes downcutting of the stream channel by abrasion, as well as headward erosion of the stream. The channel is thus deepened and lengthened, enlarging its drainage basin and increasing topographic relief. Downcutting also widens the channel by undermining its sides, causing them to collapse into the moving water (mass movement) and be transported downstream. The erosion process at first increases local relief, and then decreases it as the stream valley widens. Neighboring streams can be at very different stages of this process, depending on factors such as local variation in types of bedrock or loose sediment, rainfall intensity, or human intervention in the form of dam construction, plowing, or spreading of concrete for highways and buildings.

Streams transport material in three ways: by traction (bed load), by suspension, and by solution. In traction transport sediment either rolls along the stream bottom or moves in a hopping fashion, called *saltation*. Moving water creates stress on the stream bottom because of friction between the water and the individual grains. This friction generates upward eddies of water strong enough to lift the grains off the bottom so they can be moved downstream. The higher the degree of stress, the larger the grain size and the greater the number of grains the stream can move. The size of the largest particle the stream can move is termed the *competence* of the stream. Competence varies as the sixth power of stream velocity. A stream's velocity is determined mostly by its *gradient*, expressed as the number of feet (or meters) the stream descends for each mile (or kilometer) along its flow path. Gradients are steepest in headwater tributaries; in mountainous areas they can exceed 250 ft/mi. The lower reaches of the Mississippi River, in contrast, have gradients of only 0.1 in./mi.

Capacity is the total amount of sediment a stream can carry. Most streams with high capacities also have high suspension loads. Usually, the capacity of a stream varies as the third power of the velocity.

Muddy sediment erodes less easily than sandy sediment because it is more cohesive. Clay flakes are shaped like pieces of paper and adhere to each other almost immediately when they make contact at the stream bottom. The adhesion makes them more difficult for the moving water to erode, or pick up. Once they are picked up, however, they are transported easily because of their small size, so a stream can transport large volumes of mud very rapidly. The amount of *suspended load* in a stream depends mostly on water discharge rather than on bottom shear stress.

Solution load is important in water-quality investigations. In unpolluted areas the ions in the water have dissolved from the rocks in the drainage basin. Waters affected by human activities, in contrast, can contain almost anything.

As a stream progresses through its lifetime of sculpting the land surface, it creates features that are very important in environmental geology. Most of these features develop during the middle part of the stream's life cycle, as the stream changes its major work from downcutting (vertical erosion) to lateral cutting (horizontal erosion). As lateral

Figure 9.5

The evolution of stream meanders results from both erosion on the outside of a curve in the stream channel, where velocity is greatest, and deposition on the inside of the curve, where velocity is lowest. (a) Streamflow is deflected by an irregularity and moves to the opposite bank, where erosion begins. (b) Once the bend begins to form, the flow of water continues to impinge on the outside curve, so a meander loop develops. At the same time, deposition occurs on the inside of the bend as a result of the lower stream velocities in that area. (c) The meander is enlarged and migrates laterally, with the contemporaneous growth of a point bar. A general downslope migration of meanders occurs as they grow larger and ultimately cut themselves off to form oxbow lakes. *Introduction to Physical Geology, 2/e by Hamblin,* ©*1994. Adapted by permission of Prentice-Hall, Inc., Upper Saddle River, NJ.*

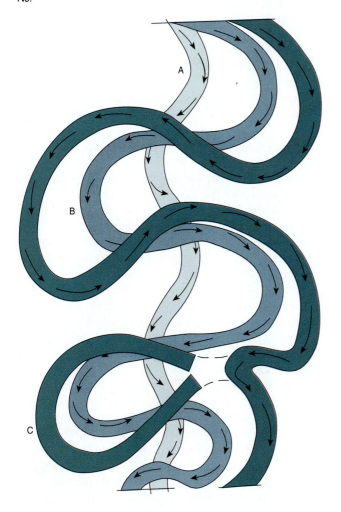

Slower, less turbulent water along the inside of the bend results in sediment deposition. Meanders thus become increasingly exaggerated, and the width of the meander belt increases. As the width increases, meanders are more easily cut off from the stream.

When a stream overflows its channel, or floods, it creates raised ridges called *levees* at the channel margins. Levees develop because water velocity decreases sharply as it leaves the channel, causing suspended sediment to be deposited almost immediately, building a ridge at the channel margin. With each succeeding flood, the height of the levee increases.

The area into which the stream spills over during floods is called its *floodplain;* this is the surface on which the meanders are located. The width of the meander belt can be no greater than the width of the floodplain and is often much smaller. Some wide-floored valleys have elevated, nearly level benches called *terraces* along their margins (Figure 9.6). Typically, these benches occur at the same elevation on both sides of the channel. Terraces are remnants of former valley floors, valley-wide floodplains that once existed at higher levels than that of the present floodplain. The renewed downcutting that left the terrace might have been caused by either climatic change or tectonic uplift in the stream's headwaters area.

Problems

1. Why is the volume of water in surface runoff always less than the volume of precipitation?

2. How do both the presence and the type of rock or sediment at the earth's surface affect the amount of runoff? Compare, for example, the effects of granite versus gravel or sandstone versus shale.

3. In Figure 9.7, assume that at point A in the river the elevation of the bottom is 70 feet, and at point B it is 65 feet. How much did the formation of the Caulk cutoff in 1937 change the gradient of the river?

4. What was the rate of movement of the meander bend from west to east between 1827 and 1846?

5. Why do the topographic contour lines in the Caulk Neck–Caulk Point region roughly parallel the meander?

Figure 9.6

Sketch of a valley filled with alluvium that has been eroded into terraces. Each terrace is coded by a pattern.

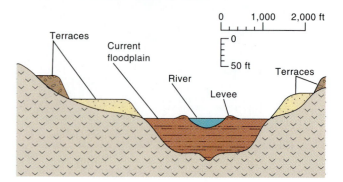

erosion proceeds, the stream channel begins to meander (Figure 9.5). With time, the *meanders* enlarge and move downstream; their rate of movement depends on water discharge and local geology. Meanders can move laterally at speeds of tens or even hundreds of meters per year, although rates below 10 m/yr are more common on smaller streams.

Meanders develop best in muddy streams—streams with a high ratio of suspension load to bed load. The meanders migrate because the deepest, swiftest, and most turbulent section of the stream channel, where erosion is most active, lies along the outer margin of each meander bend.

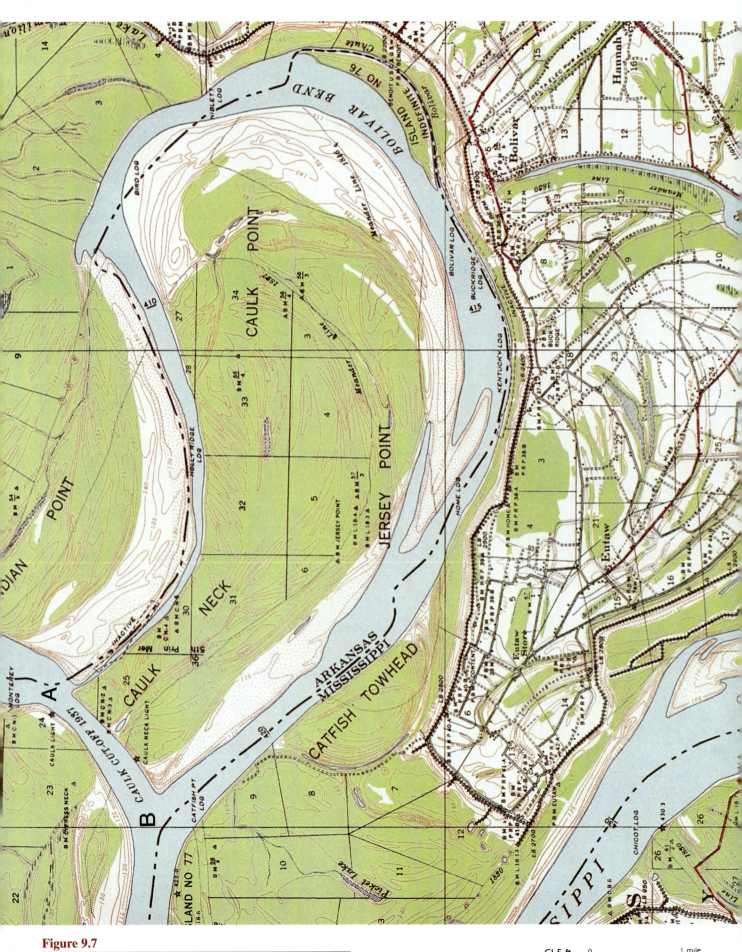

Figure 9.7

Topographic map of a part of the Lamont, Arkansas Quadrangle.

CI 5 ft   0   1 mile

6. What is the maximum local relief of the natural levees in this area?

7. The topographic map shows that the state boundary between Arkansas and Mississippi was drawn in the mid 1800s. By examining the map you can determine the approximate date when the state boundary was established between Arkansas and Mississippi. Estimate this date. What are the pros and cons of defining state boundaries that follow the course of a river?

8. Interference in natural processes by humans can greatly alter erosion rates and the size and shape of stream channels. Figure 9.8 shows the general relationship between the use to which land is put, the condition of stream channels on the land, and the amount of sediment carried by the stream. Note how these variables have changed between 1780 and 1980. Explain

   a. the changes in sediment yield (erosion) as land use changed.

   b. the condition of the stream channel as land use changed.

Further Reading/References

Leopold, Luna B., 1968. *Hydrology for Urban Land Planning: A Guidebook on the Effects of Urban Land Use.* U.S. Geological Survey Circular 559, 18 pp.

Leopold, Luna B., 1994. *A View of the River.* Cambridge, Massachusetts, Harvard University Press, 298 pp.

Watson, Ian, and Burnett, Alister D., 1993. *Hydrology: An Environmental Approach.* Ft. Lauderdale, Florida, Buchanan Books, 702 pp.

Wolman, M. Gordon, and Riggs, H. C. eds., 1990. *Surface Water Hydrology. The Geology of North America,* v. O-1, Boulder, Colorado, The Geological Society of America, 374 pp.

Figure 9.8

Variation in sediment yield through time in the Appalachian foothills of eastern United States. *From M. G. Wolman, "A cycle of sedimentation and erosion in urban river channels,"* in Geografiska Annaler, *[49A, pp. 385–396, 1967]. Copyright © 1967 Scandinavian University Press, Oslo, Norway. Reprinted by permission.*

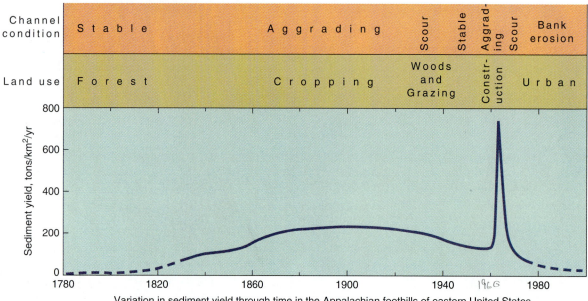

Variation in sediment yield through time in the Appalachian foothills of eastern United States.

E X E R C I S E 10

FLOODS

Floods are the most common of the many geologic catastrophes that plague humankind; they affect more people than all other natural hazards combined. As much as 90% of the damage related to natural disasters (excluding droughts) is caused by floods, at an estimated annual cost of $2.4 billion. About 7% of the land area of the 48 conterminous United States is subject to flooding, and these floods can cover hundreds of thousands of square miles (Figure 10.1). More than 20,000 communities, with over 6 million single-family homes and representing perhaps 10% of the total U.S. population, are located on flood-prone land. Of the recent major disasters declared by U.S. presidents, 85% were associated with floods.

Floods can result from several unusual events. The hurricanes that occur each year along the Atlantic and Gulf coasts produce enormous surges of ocean water that can drown nearshore areas. In fact, most of the damage from hurricanes results from these surges and associated rainfall. The greatest of these hurricane floods occur along the fringe of the Gulf of Mexico, but many smaller but damaging examples occur along the Atlantic coast.

Many floods have a common cause: too much rainfall within a short time. When rain falls slowly, enough water is absorbed by soil and bedrock or channelled into stream courses to prevent inundation of the surrounding area. If rainfall is excessive, however, flooding occurs (Figure 10.2), often with loss of life and extensive property damage. Flooding can also occur if heavy snowfall melts quickly in abnormally warm spring weather. Other floods are secondary, resulting perhaps from breaching of a natural dam

formed by a landslide across a stream, or from human disruption of the natural environment, as in the case of a dam failure. Flood control is a major environmental problem in many areas. In the United States an important area of concern is the lower Mississippi River. Overseas, the most continually endangered region might be the nation of Bangladesh, whose 130 million people live on 55,000 square miles of low-lying, perennially flood-prone land along the Ganges River.

The key controls over an area's tendency to flood are (1) the amount and distribution of precipitation and (2) the topographic and geologic characteristics of the rainfall catchment area. Average annual rainfall ranges from virtually zero in the driest deserts to 451 inches at one location in Hawaii. Perhaps more important than such averages, however, is the seasonal distribution of the precipitation. Clearly, if all the rain falls within a one-month period, flooding is more likely than if it falls fairly evenly throughout the year. Meteorologists understand the general controls that govern the temporal distribution of precipitation, but most flooding problems are caused by unusually heavy rainstorms that occur at irregular intervals.

Topography has a direct influence on flooding. The most areally extensive floods (though not necessarily the most damaging) occur in low-lying, downstream areas. The downstream areas have larger catchments for collecting precipitation, but shallower stream channels for containing it. A *floodway* consists of the area of channel and immediately surrounding flat ground that provide the avenue for flood waters (Figure 10.3). In inhabited areas humans have important

Figure 10.1a

Map showing distribution of great floods in the conterminous United States since 1889. *Source: U.S. Geological Survey.*

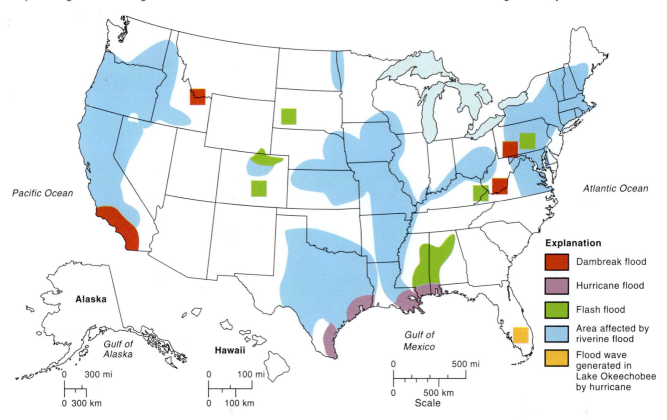

Explanation

- Dambreak flood
- Hurricane flood
- Flash flood
- Area affected by riverine flood
- Flood wave generated in Lake Okeechobee by hurricane

Figure 10.1b

Curves showing the approximate limits of the largest floods experienced in the United States at successive times. The flattening of the curves at larger drainage areas indicates that the peak discharges per unit of drainage area are smaller as the size of the drainage basin increases. With the passage of time, the curves have moved up as larger floods have occurred. The longer the period for which records are available, the more likely it is that a flood will occur with an unusually large discharge. *Source: U.S. Geological Survey.*

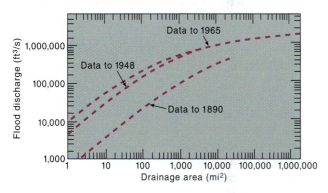

Figure 10.2

Flood frequency curve for Eel River in Scotia, California, based on data collected from 1932 to 1959. The graph shows how often on average a given discharge will occur. *From Gary B. Griggs and John A. Gilchrist,* Geologic Hazards, Resources, and Environmental Planning, *2d ed. Copyright © 1983 Wadsworth Publishing Company, Belmont, CA. Reprinted by permission.*

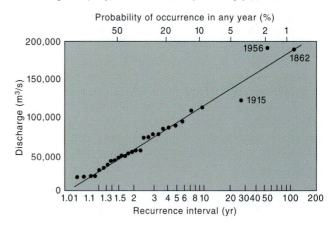

effects on the amount of runoff. By devegetating, bulldozing, building on, or otherwise changing the land surface, humans greatly affect the amount of water that infiltrates into the ground or runs off the surface. For example, clearing a forest increases the intensity with which rain hits the soil, reduces infiltration, and increases surface runoff into streams. Structures such as houses and pavement increase runoff by decreasing the natural surface available to soak up precipitation, and thereby increase the frequency of flooding. In 1974 the United States Geological Survey evaluated the extent and development of urban floodplains. Among the 26 moderate to large cities studied, an average of 52.8% of their total floodplain areas had been urbanized. Values such as 97% for Great Falls, Montana; 89.2% for Phoenix, Arizona; 83.9% for

Figure 10.3

Perspective sketch showing relationships of the river channel to bottomland. The cross-section shows flood stages and flood frequencies. *From Moss and others, 1978.*

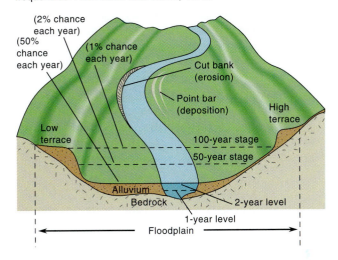

Tallahassee, Florida; and 83.5% for Harrisburg, Pennsylvania, indicate why so many urban areas are susceptible to yearly flooding and its associated damage.

Whether a stream will overflow its banks, spread over its floodplain (which becomes the floodway), and perhaps spread farther into a neighboring town, depends on the stream's discharge. During extremely high floods, city streets may become part of the floodway. A stream's discharge is the amount of water carried by the stream. It depends on the size of the stream channel (and floodway during floods) and the velocity of the stream. Discharge is calculated from the equation

$$Q = AV,$$

where Q = discharge (ft/s)

    A = cross-sectional area of the stream channel at the location where discharge is determined (stream width × stream depth; ft)

    V = average velocity at the site (ft/s)

The amount and character of the debris carried in a flood is affected not only by the intensity of the precipitation but also by the nature of the bedrock. For example, a watershed underlain by poorly consolidated, fine-grained sedimentary rocks produces debris consisting of mud, while a watershed composed largely of harder rock yields coarser debris. Overland flows move these materials and transport them to stream channels. On steep slopes with loose, broken rock or poorly consolidated soils, downpours can trigger avalanches, slides, or mudflows. Structures located in the path of such a moving mass are often damaged by the impact. Solid materials derived from watershed slopes lodge in channels, later to be swept away by currents, usually during floods.

FLOOD-FREQUENCY ANALYSIS

The objective of a flood-frequency analysis is to determine how often, on the average, a particular region can expect a flood of a certain magnitude. The steps in the analysis are as follows:

1. Obtain the streamflow records of a particular gauging station for all the years during which records have been kept. Choose the station with the longest, most complete record.
2. Identify and list the highest discharge rate for each year.
3. Rank the water discharges in decreasing order.
4. To determine the recurrence interval for each discharge, use the formula

$$\text{Average Recurrence Interval} = \frac{n + 1}{m}$$

where n = number of years of record

    m = rank or position of any individual discharge in the series

5. Using these values, construct a graph on semilog graph paper of recurrence interval versus discharge (or stage). Connect the data points with a line of best fit.

It is best to view flood hazard statistics in terms of probability rather than recurrence interval (Table 10.1). One value is actually the reciprocal of the other. A 50-year flood, for example, has a 1/50 or 2% chance of occurring in any given year; a 25-year flood has a 4% chance. Always keep in mind that flood predictions are only probabilities. Two 50-year floods can occur in successive years, followed by two centuries during which no 50-year floods occur at all. The predictive graph gives no guarantees for any particular year; rather, it offers only averages based on past occurrences.

Few streams have records spanning 100 years, so we must often extend plots beyond the data to estimate the magnitude of 100-year events and locate the 100-year floodway. The longer the extrapolation, the more uncertain the estimate. Another factor that leads to uncertainty in flood-frequency curves and recurrence-interval computations is the occurrence of a very large flood (say the 100-year event), within a short period of record (for instance, 20 years); it diverges from all the other data points.

FLOOD PREVENTION AND PROTECTION

Many communities in flood-prone areas near rivers try to prevent floods from inundating their town. There are several ways to do this. Each method has its benefits and limitations, and all methods are expensive. The more protection desired, the greater the expense. The most common level of protection used is protection against the ravages of a 100-year flood.

One method is floodplain zoning, restricting construction in the area alongside the river that would be submerged

TABLE 10.1

Likelihood of Floods of Different Magnitudes. A 100-year flood has a 1% chance of occurring in any specific year, a 9.6% chance during a 10-year period, a 22% chance during a 25-year period, etc.

One Hundred Years	Fifty Years	Twenty-five Years	Ten Years	Any One Year	Return Period, Years
Chance (%) of at Least One Flood of at Least This Size in a Certain Number of Years					
				50	2
				40	
				30	
				25	
				20	5
		99	80	15	
	99.9	94	65	10	10
	90.5	71	40	5	20
86	63	40	18	2	50
63	39	22	9.6	1	100
39	22	12	5	0.5	200
18	9.5	5	2	0.2	500
9.5	4.8	2.5	1	0.1	1000
5	2.3	1.2	0.5	.05	2000
2	1.0	0.5	0.2	.02	5000
	0.5	.25	0.1	.01	10,000

From B. M. Reich, Water Resources Bulletin, 9:187, 1973. Copyright © 1973 American Water Resources Association, Bethesda, MD. Reprinted by permission.

during a 100-year flood. Knowing the frequency of different discharges and the topography of the area around the river, the extent of flooding during a 100-year flood can be determined. Zoning ordinances of any kind are contentious, and zoning for an event expected, on average, to occur only once in a hundred years is even more so.

Another method of flood prevention is *channelization,* improving the stream's channel so it can hold larger discharges without overflowing. This can be accomplished by dredging the channel so larger discharges can be accommodated. The channel may also be lined with concrete to keep the river from meandering through the town. Channelization may also be accomplished by building higher levees than the stream has created naturally.

All methods of channelization have serious side-effects. They increase flooding downstream so that what used to be a 100-year flood for the folks downstream is now a 50-year occurrence, or perhaps a 25-year occurrence. Channelization also increases erosion downstream and affects the ecology both upstream and downstream.

A third way to protect against floods is by building dams across the river to contain floodwaters. However, the lake or reservoir that forms behind a dam tends to fill with sediment carried by the inflowing stream. This water-filled, topographically low area behind the dam traps sand and mud that was once carried downstream—sediment that can now be removed only by draining and dredging the lake, a very expensive process.

FLOOD-LOSS REDUCTION

Several steps can be taken to decrease the effects of flooding on human populations. The most important is to educate the public about the frequency and dangers of local flooding. Without an effective educational program, people will not take the steps necessary to ensure safety and minimum loss of life and property.

Many governments have passed regulations concerning floodplain use and occupancy. Foremost among these are zoning laws that limit or regulate the types of construction permitted in specific areas adjacent to stream courses. Such legislation generally requires agreement between real estate developers who own the land and want to maximize their profits from it, and the city or state government, whose main concerns might be different. Zoning laws often are accompanied by building codes, or construction standards that include floodproofing requirements such as placing shields around buildings, or erecting buildings on stilts that raise the bottom floor to several feet above ground level.

Many insurance companies sell flood insurance, and a community can opt to require its purchase by those who insist on owning structures in flood-prone areas. Some communities offer tax incentives for investing in ways to reduce flood loss. However, many Americans dislike programs that force individuals to protect themselves against flooding. Instead, they believe the government's responsibility lies in providing relief programs or subsidies, such as low-interest loans, if a catastrophe occurs. Numerous disaster-relief programs now exist at both state and federal levels.

Problems

1. In 1979 Houston, Texas, had three 100-year floods. What does this do to your confidence about building your house near a river?

2. Based on the changes of the bounding curve (Figure 10.1b) in the graph of the largest U.S. flood discharges versus drainage basin area, do you think the position of the most recent curve is likely to change significantly if more recent data (post-1965) are considered? Explain.

3. The accompanying table (Table 10.2) shows mean annual discharges for the Chikaskia River near Blackwell,

TABLE 10.2

Mean Annual Discharges for the Chikaskia River near Blackwell, Oklahoma, for 1937–1974

Rank	Year	Mean Discharge (ft³/s)	Recurrence Interval (years)
	1937	266	
	1938	419	
	1939	159	
	1940	76.6	
	1941	199	
	1942	690	
	1943	282	
	1944	694	
	1945	884	
	1946	228	
	1947	687	
	1948	728	
	1949	1170	
	1950	375	
	1951	1450	
	1952	254	
	1953	109	
	1954	71	
	1955	307	
	1956	168	
	1957	979	
	1958	420	
	1959	468	
	1960	908	
	1961	656	
	1962	535	
	1963	184	
	1964	189	
	1965	966	
	1966	97.6	
	1967	158	
	1968	337	
	1969	661	
	1970	439	
	1971	161	
	1972	151	
	1973	1130	
	1974	962	

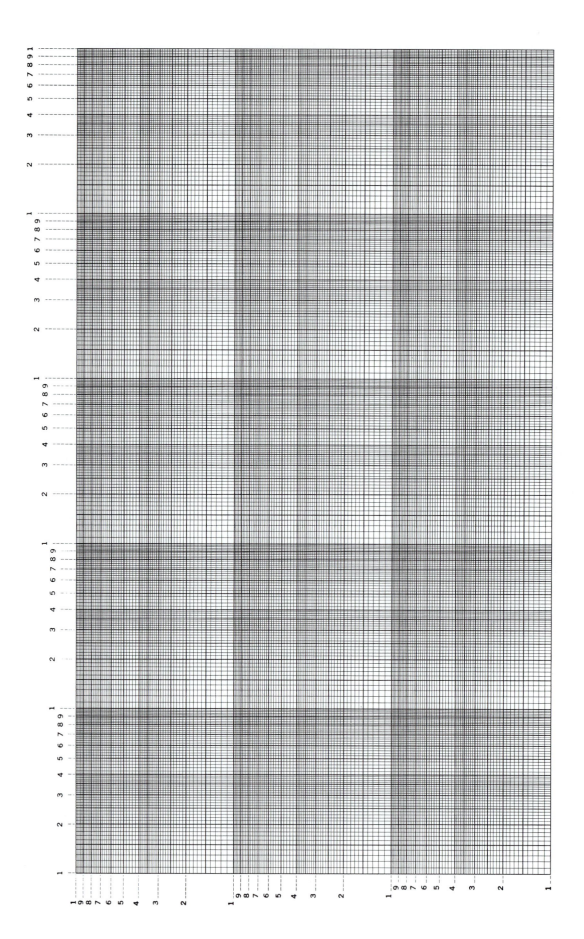

Oklahoma, over a 38-year period. Rank the discharges and calculate the recurrence interval for each year, then plot the data on the log-log graph paper provided. Eyeball a best-fit line through the 38 data points.

a. Compare the discharge indicated by your line with that determined by your two nearest neighbors in class, with respect to the 2- and 100-year recurrence intervals. Does your agreement differ for the 2-year versus the 100-year estimates? Would you expect a difference? Why or why not?

b. Suppose you had plotted the highest 1-month mean discharges for each year rather than the mean discharge for each 12-month period. Do you think the line for the 1-month data would be above or below the mean for each year? Explain. Suppose you plotted the highest 1-day discharges? One-hour discharges? What conclusions do you draw from thinking about these different approaches?

4. Black Bear Creek flows west to east through the center of the Garber Quadrangle in central Oklahoma (see fold-out map at back of book). After months of searching you find a site on which to build your dream house, and the current owner is willing to sell the property at a price you can afford. The area is the SE 1/4 of section 20, T22N R3W. However, as an environmentally knowledgeable person you are concerned about the possibility of flooding. You contact the State Water Resources Board and obtain the rating curve (the relationship between stream discharge and river stage), as well as discharge recurrence data for the creek at a gaging station alongside the property (Figures 10.4, 10.5).

a. Outline on the fold-out map (Figure 10.6 at the back of this book) the drainage basin of Black Bear Creek.

Figure 10.4

Discharge versus river height for Black Bear Creek in section 20, T22N R3W, Garfield County, Garber Quadrangle, Oklahoma. *Source: Data from Oklahoma Geological Survey.*

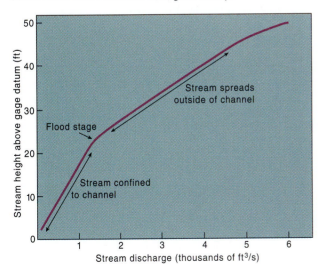

Figure 10.5

Recurrence interval versus stream discharge for Black Bear Creek in the center of section 20, T22N R3W, Garfield County, Garber Quadrangle, Oklahoma. *Source: Data from Oklahoma Geological Survey.*

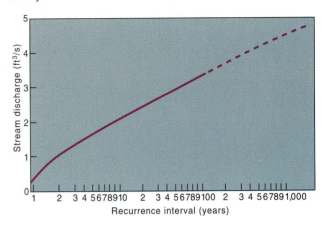

How many square miles of the drainage basin of Black Bear Creek are present upstream from the property?

b. How frequently can you expect the creek level to reach your front door, which will be at an elevation of 1,050 feet?

c. Are you willing to build your house there in light of your answer? What frequency of flooding is acceptable to you?

d. Would (or should) your decision about house construction be affected by whether or not a flood has actually attained the 1,050-ft level recently?

e. Should the minor stream tributary on the property be of concern to you? Why or why not?

f. Another potential homesite is the N 1/2 of section 22, T22N R4W. Give two reasons why this might be a better choice from a hydrologic viewpoint.

5. Suppose you are the environmental geologist called to choose the best site for constructing a flood-control dam on a large river. List the factors you think important in selecting the location, briefly explaining the significance of each factor. Consider the lithology and coherence of the rocks around the potential site, the position of the site in the drainage basin, and any other factors you consider potentially significant.

6. Your small local reservoir is filling with mud and consideration is being given to dredging it to increase its capacity. The stream entering the reservoir drains an area of 100 square miles and carries a yearly sediment load of 1,000,000 pounds of mud per square mile. The mud weighs 150 pounds per cubic foot. The reservoir is 50 years old. Dredging costs 10 cents per cubic foot. How much would it cost to dredge the reservoir of its mud? How much should the town set aside each year for dredging?

Further Reading/References

Cudworth, A. G., Jr., 1989. *Flood Hydrology Manual.* Denver, Bureau of Reclamation, U.S. Dept. of the Interior, 243 pp.

Gross, E. M., 1991. "The hurricane dilemma in the United States." *Episodes,* v. 14, pp. 36–45.

Gruntfest, E., and Huber, C. J., 1991. "Toward a comprehensive national assessment of flash flooding in the United States." *Episodes,* v. 14, pp. 26–35.

Leopold, L. B., 1968. *Hydrology for Urban Land Planning.* U.S. Geological Survey Circular 554, 18 pp.

Wahlstrom, E. E., 1974. *Dams, Dam Foundations, and Reservoir Sites.* New York, Elsevier, 278 pp.

Williams, G. P., and Wolman, M. G., 1984. *Downstream Effects of Dams on Alluvial Rivers.* U.S. Geological Survey Professional Paper 1286, 83 pp.

E X E R C I S E 11

GROUNDWATER

Groundwater is one of our most important natural resources, currently supplying 20–25% of the water used in the United States. Subsurface water can occur in any porous sediment or rock, but amounts large enough for intensive exploitation are found mostly in sedimentary materials. These materials underlie about 65% of the world's land surface, making groundwater a widely distributed natural resource. In most areas of sedimentary rocks, it is easier to drill a well and find water than to drill and not find it. This fact explains most of the successes attributed to *water witchers*. Drilling in areas of igneous and metamorphic rocks, however, produces groundwater only in some cases in which the rocks are fractured or weathered. Lava flows commonly crack and fragment at their upper surfaces because of rapid cooling; when they are buried beneath younger materials, they retain these pores and can become important water sources, as they are in several northwestern states and Hawaii.

Sediments and rocks that yield water in amounts large enough to be significant to humans are called *aquifers* (Figure 11.1). Some aquifers are unconfined, meaning that the water-bearing unit extends up to the ground surface; others are capped by an impermeable confining layer, called an *aquiclude,* that does not transmit fluid. Confined aquifers occur at depth, unconfined aquifers near the ground surface. Aquifers, whether confined or unconfined, have a level below which the pores are full of water and above which they contain mostly air mixed with small amounts of water held between adjacent grains by capillary forces. The surface or narrow zone between these two regions is called the *water table.* Wells drilled for water production must penetrate to beneath the water table.

Environmental geologists are interested in (1) depth to the water table; (2) direction and velocity of flow and discharge of the aquifer; (3) replenishment rate and storage capacity of the aquifer; and (4) the sources, types, and movement of possible pollutants.

Water enters the ground from precipitation, lakes, and streams, filtering downward through permeable materials until it reaches the water table. The water table can occur at any level between the basal aquiclude and the ground surface. If a subsurface water table intersects a permeable fracture that extends upward to the ground surface, a *spring* forms (Figure 11.2). Springs also form where the water table intersects a hillside, making them common in mountainous areas. Contrary to advertising claims, this spring water is no more or less healthful than unpolluted water from other sources. The level of the water table is determined by the balance between inflow rate to the aquifer and the rate of withdrawal or discharge. Generally, an unconfined water table has an irregular surface shaped like a subdued replica of the overlying topography. The high areas of the water table are groundwater divides, the equivalent of aboveground topographic divides. Both above and below ground, water flows away from the divides. Drier climates have deeper groundwater tables and less pronounced similarity between surface topography and the water table.

Figure 11.1

Cross-section showing confined and unconfined aquifers. The potentiometric surface is the level to which the groundwater rises without being pumped. The cone of depression reflects a rate of pumping greater than the water's rate of replenishment.

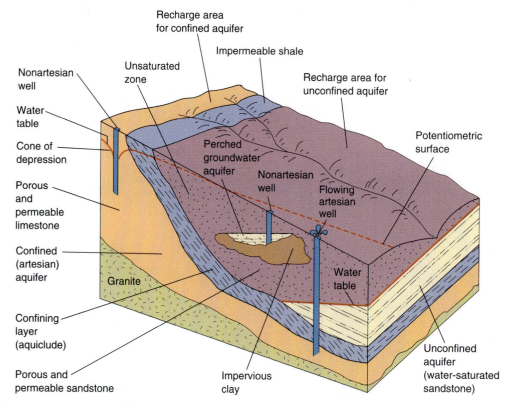

GROUNDWATER MOVEMENT

Typically, water wells drilled into confined aquifers are drilled where the rock is deeper than the recharge site, which is usually the ground surface. The groundwater at depth is under pressure from the water updip in the aquifer. As a result, when the drill pierces the aquifer, the water rises above the base of the confining bed. A well in which this rise occurs is an *artesian well*. Sometimes the pressure is great enough to make the water rise above ground level, but in most wells the water rises only partway to the surface and must be pumped the rest of the way. Free-flowing water wells were more common in the past (Figure 11.3), before intensive subsurface-water use lowered regional water levels and reduced the pressure head in confined aquifers. Water yields from aquifers range from a few gallons per hour to perhaps 20 gallons per second in very permeable rocks with high hydraulic gradients.

Several different types of maps are used to show variations in groundwater movement over an area. Contour maps can show variables such as hydraulic gradient, water-table height, potentiometric surface, flow velocity, or changes in water-table height as water is withdrawn from an aquifer. Such maps can be compared to maps of aquifer thickness (isopach maps), lithologic characteristics, or permeability. Each type of map supplies a different kind of insight into the variables that control the availability of subsurface water.

AVAILABILITY OF GROUNDWATER

Not all precipitation is discharged at the surface as runoff; some percolates into soil, loose sediment, or rock, and can descend many hundreds of feet into the ground. This groundwater is a common source of water for human consumption. The controls on the amount of infiltration are partly climatic (How much rain? How intense? For how long?), partly topographic (Are slopes steep or gentle? Is the rain ponded or free to flow downslope?), and partly lithologic. The lithologic controls involve the type of surface material, its porosity, and its permeability.

POROSITY

Porosity refers to the amount of void space between grains:

$$\% \text{ porosity} = \frac{\text{pore volume}}{\text{rock volume}} \times 100$$

The main controls of porosity in sedimentary materials are the distribution of grain sizes and the amount of chemically precipitated cement among the particles. When there is a wide range in particle sizes, the smaller grains lodge between the larger ones, reducing the amount of pore space. Rocks that consist mostly of clay minerals, such as shales, have no porosity. The sheet-like clay flakes are flexible and compact tightly so that the pore space originally present is eliminated.

Figure 11.2

Geological factors in the location of springs. *From S. N. Davis and R. J. M. DeWiest,* Hydrogeology. *Copyright © 1966 by John Wiley & Sons, Inc., New York, NY. Reprinted by permission of John Wiley & Sons, Inc.*

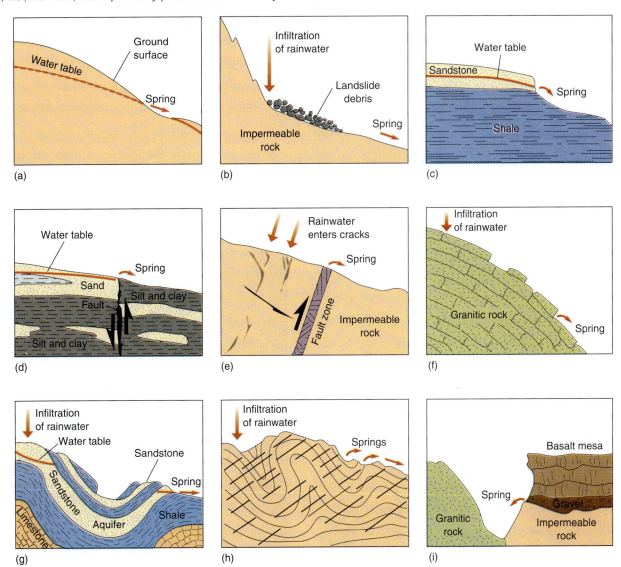

Chemically precipitated cements that form after burial of the sediment and bind the grains together into a rock do so by filling pore spaces. Sometimes all the pores are filled and porosity is reduced to zero. Sometimes the pores are only partly filled so that some pore space remains. A fragmental rock can contain water or oil only if cementation was incomplete. When deposited, accumulations of gravel or sand have porosities of 30–40%. Sandstones from which we get well water or petroleum usually have porosities of 5–15%.

Pore spaces in sedimentary rocks are located among the grains. As a result the pore spaces are small, normally smaller than the diameters of the grains in the rock. The fragmental grains in an average sandstone are about 0.3 mm (0.01 in.) in size and the pore sizes are even smaller. But despite their small size, pores are numerous enough to hold a very large amount of fluid. The number of pores can be as great as the number of grains, and the number of grains in even a small pile of sand is very large. Consider, for example, a sandstone layer 10 feet thick that extends over an area of 1 square mile. The volume of rock is 278,784,000 ft^3 (5,280 × 5,280 × 10). If it contains 10% pore space, the volume of pore space is 27,878,400 ft^3. One gallon of water occupies 0.134 ft^3, so the pores in this layer of sandstone can hold more than 200 million gallons of water. The average American uses about 500 gallons per day for household purposes, so 200 million gallons of water would supply the household needs of a city of 400,000 for one day.

Figure 11.3

Photograph of one of the highest artesian well flows on record, drilled in 1909 into the Dakota Formation (Cretaceous) at a depth of about 775 feet in Woonsocket, South Dakota. The well was 6 inches in diameter and had a flow of 1,150 gallons per minute, and the water spout rose more than 100 feet above the ground surface.

PERMEABILITY

Permeability refers to the ease with which a fluid flows through a rock. It depends on the size of the pores and the viscosity of the fluid. Larger pores and less viscous fluids make for easier (faster) flow. Hence, sandstones with a wide range in grain size have lower permeabilities than sandstones with a narrower range in grain size. When the smaller grains lodge between coarser ones, the pore diameters are reduced. Most destructive to permeability are clay minerals. Because of its sheetlike shape, a clay flake has a very large surface area for the small volume it occupies, and therefore greatly increases the amount of surface over which a fluid must flow as it moves through a rock. This increase in "fric-

tion" (actually electrostatic attraction) between the fluid and the grain surfaces reduces the flow velocity of the fluid. A few percent of clay in a sandstone is enough to virtually eliminate the flow of water or petroleum through an otherwise productive layer.

Mudrocks cannot transmit fluids unless fractured. The average mudrock contains about 60% clay minerals, which bend to conform to each other's shape when compacted. Shales are a common aquiclude.

Limestones are composed entirely of calcium carbonate, which is fairly easily dissolved (soluble) in underground water (Figure 11.4). Sometimes the dissolution extends through a very large volume, producing a cavern of considerable size.

Figure 11.4a

Partially interconnected (in the third dimension) solution cavities in a drill core of microcrystalline limestone. Width of core is 2.5 inches.

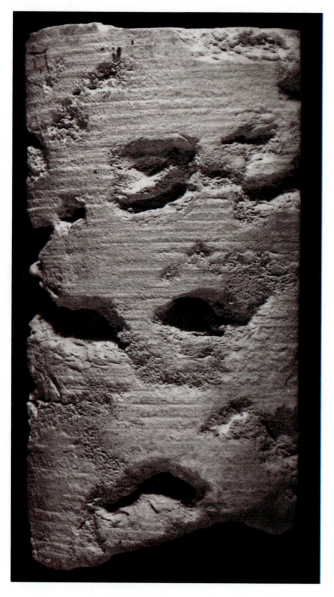

Figure 11.4b

Extremely large "pores" in limestone, Carlsbad Cavern, New Mexico. The pore in which the man is standing bifurcates into two smaller ones in the direction he is facing. Limestone caverns are simply very large pore networks formed by dissolution of soluble calcium carbonate.

Carlsbad Cavern in New Mexico and Mammoth Cave in Kentucky are examples of huge dissolution cavities in limestones.

HOW MUCH WATER CAN AN AQUIFER PRODUCE?

The permeability of a rock is essential information for an environmental geologist interested in the amount of water or oil that can be produced by a sedimentary rock. Permeability to water is defined mathematically by an algebraic relationship first recognized in 1856 by the French hydrologist Henri Darcy.

$$\frac{\text{Water}}{\text{Discharge}} = \text{permeability} \times \frac{\text{slope of the}}{\text{water table}} \times \frac{\text{cross-sectional}}{\text{area of the aquifer}}$$

The slope of the water table is obtained by dividing h, the difference in height of the water table at two points along the direction of flow, by l, the horizontal distance between the two points. Written symbolically, Darcy's law is

$$Q = K \frac{\Delta h}{\Delta l} A$$

Darcy's formula allows the environmental geologist to calculate the amount of water that will flow through the aquifer (be available in a water well) each day. For example, suppose a confined aquifer is 60 feet thick and 5 miles wide. Wells have been drilled 1 mile apart in the direction of flow. The level of water in one well is 150 feet and in the other is 140 feet, which defines the slope of the water table. The permeability of the aquifer is 4.5 feet per day. How much water will flow through the aquifer each day?

$$Q = K \frac{\Delta h}{\Delta l} A \text{ (which is aquifer width} \times \text{aquifer thickness)}$$

Plugging in the numbers,

$$Q = (4.5 \text{ ft/day}) \frac{150 \text{ ft} - 140 \text{ ft}}{5,280 \text{ ft}} (60 \text{ ft})(26,400 \text{ ft})$$

$$= 13,600 \text{ ft}^3 \text{ per day}$$

This is 104,743 gallons. The average American uses about 500 gallons per day for bathing, drinking, and cooking, so 100,000 gallons is enough for about 50 families.

THE EFFECT OF OVERPRODUCTION

If water is continually withdrawn from an aquifer faster than it can be replaced at the recharge area, two things may happen.

1. The level of the water in the well will decline and eventually the well will have to be abandoned (Figure 11.5).
2. Salt water will infiltrate the aquifer and eventually make the well water unfit to drink (Figures 11.6, 11.7).

SALTWATER ENCROACHMENT

Most of the sedimentary rocks in the geologic column were deposited in the oceans, so their original pore water was sea water. Subsequent flushing by surface water often has caused freshening at shallow depths, so that the salinity of pore waters generally increases with depth. Potable waters (those safe and palatable for human use) can occur to depths of perhaps 2,000 feet, but most water wells are limited to depths of about 500 feet because of drilling costs.

The principles by which water occurs in and moves through porous media apply to saline as well as fresh water. Encroachment of salt water into a freshwater aquifer can be induced in many ways, such as by pumping too much fresh water from wells; by puncturing the aquicludes that protect freshwater aquifers (by wells, tunnels, dredging, or other construction); by ponding or otherwise enabling saline water to move downward into freshwater aquifers; or by discharging saline wastes directly into aquifers. Salty water, including sea water, is denser than fresh water and thus tends to move inland under it, creating a freshwater-saltwater interface (Figure 11.6).

An example of saltwater encroachment is found in the water wells in Union Beach, New Jersey (Figure 11.7). Until the early 1960s the chloride-ion concentration, a commonly used measure of the presence of sea water, was static at about 2 mg/l despite lowering of the water table by

Figure 11.5

Hydrograph showing changes of water level in a water well in a rural community that was founded in 1880. The well is nearing the end of its useful life.

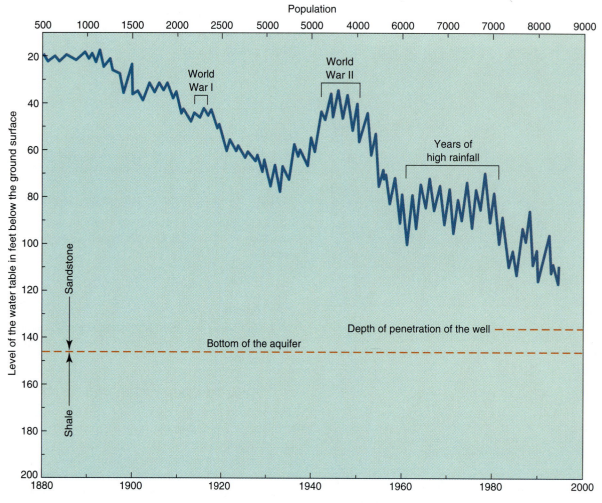

Figure 11.6

Encroachment of salt water into the lens of fresh water as a result of pumping and removal of fresh water near a shoreline. *Source: R. C. Heath, 1983, U.S. Geological Survey Water-Supply Paper 2220.*

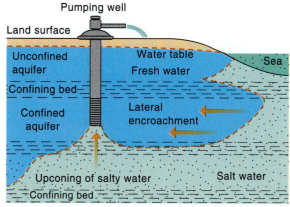

Two aspects of saltwater encroachment

Figure 11.7

Chloride concentrations in water samples from the Union Beach Borough well field, 1950–1977. *Source: F. L. Schaefer and R. L. Walker, 1981, U.S. Geological Survey Water-Supply Paper 2184.*

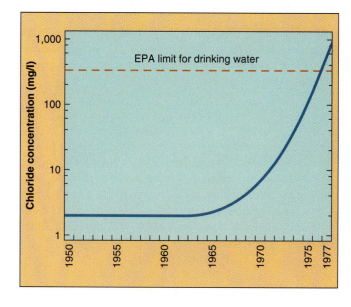

15 feet caused by withdrawals from wells. But as water use continued to increase and the water table continued to drop below sea level, the hydraulic head of the fresh water became inadequate to keep the sea water from infiltrating the aquifer. By 1977 the chloride concentration reached 660 mg/l (sea water contains 19,000 mg/l chloride ion).

GROUNDWATER AND CAVERNS

Although groundwater can move through sandy sediments at rates of several feet per day, most moves through the pore spaces of rocks very slowly. A typical velocity might be a few inches per year. Such a slow-moving fluid cannot cause mechanical erosion on the pore walls, but it can cause chemical erosion by dissolving the pore walls (the grains of the rock) if it is undersaturated with respect to the mineral forming the walls. Such is the case with shallow groundwaters and calcite. Fresh water percolating downward through soils becomes enriched in CO_2 from the decomposition of soil organic matter, and as the water enters underlying limestone, this high CO_2 level makes it undersaturated with calcite. Over thousands of years, the acidity (pH ~5) can dissolve large holes. When the holes are large enough for humans to tour, they are called caverns, for example Carlsbad Cavern and Mammoth Cave. The largest room at Carlsbad is more than 4,000 feet long, 600 feet wide, and 350 feet high—a volume of nearly one billion cubic feet, testifying to the chemical aggressiveness of shallow groundwater in limestone. The resultant water is very "hard" (>75 ppm $Ca^{+2} + Mg^{+2}$) and is not of the best quality for washing clothing or people.

The creation of large holes within a few hundred feet of the ground surface typically causes overlying rock layers to collapse, particularly when the rock contains vertical fractures (joints), as most limestones do. Such a roof-collapse is disastrous for people who live nearby (Figure 11.8).

Figure 11.8

Aerial view of a large sinkhole that formed in Winter Park, Florida, in May, 1981. Several buildings have partially collapsed into the hole, which may enlarge further in the future.

Problems

1. Suppose the water table is at the ground surface in a topographically flat area. What would be the result, that is, how would you recognize such an occurrence?

2. Suppose a stream is located above an unconfined aquifer 100 ft below it. What will be the effect on streamflow?

3. Farmer Smith is having a water-well drilled. His instructions to the drillers are to save money by stopping as soon as they encounter water. Is this a smart decision? Explain.

4. Nassau and Suffolk counties are densely populated areas on Long Island, New York. Shown are a generalized geologic cross-section of the Long Island aquifer system in Nassau County (Figure 11.9a) and the elevation of the water table on Long Island (Figure 11.9b).
   a. How many different aquifers are present on Long Island? Name them.
   b. Which aquifers are confined? Unconfined?
   c. Is the freshwater supply in any of the aquifers interconnected?
   d. In which of the aquifers do you think saltwater intrusion might be a serious problem? Explain.
   e. Which aquifer would be most vulnerable to pollution from human activities?
   f. Draw a line to locate the groundwater divide on Long Island (Figure 11.9b). Locate the divide on cross-section 11.9a. In which aquifer is the divide most effective?
   g. The thickness of the saturated zone in the aquifers shows a consistent increase from north to south on Long Island, to a maximum of about 1,000 ft at the south shore. Why does this increase occur? (Hint: examine Figure 11.9a).
   h. Which aquifer should suffer most from saltwater intrusion along the south shore of Long Island?
   i. The flow velocity of water in the Magothy aquifer varies irregularly on Long Island. What can you infer about the lithologic character of this aquifer?

5. In the map on page 84, olive lines show topography, while dashed lines signify elevation above sea level of the potentiometric surface of a confined aquifer.
   a. Outline the areas where wells will flow at the land surface without being pumped.
   b. Locate the best spot for drilling a water well, based on the relationship between topography and potentiometric surface.
   c. How far above the ground surface will the water spout at your well site?
   d. Locate the spot where a well would require the most pumping. How high would the water need to be pumped?
   e. In which direction is the recharge area for this aquifer?
   f. The potentiometric contours decrease uniformly from NE to SW. Why do you think these elevations decrease? What does this pattern tell you about the uniformity of the rock orientation of the aquifer layer in the area?

Figure 11.9a–b

(a) Generalized geologic section of the Long Island aquifer system in Nassau County. (b) Water-table altitude in Nassau and Suffolk Counties. *Source: U.S. Geological Survey Water-Resources Investigations Report 86-4141 and Professional Paper 627-E, 800 C.*

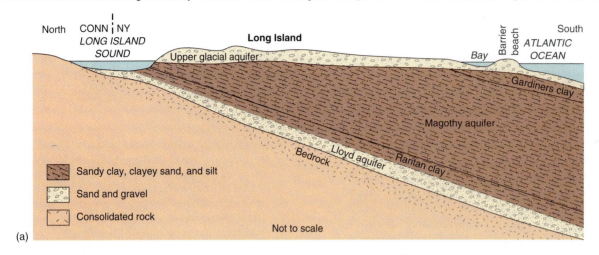

(a)

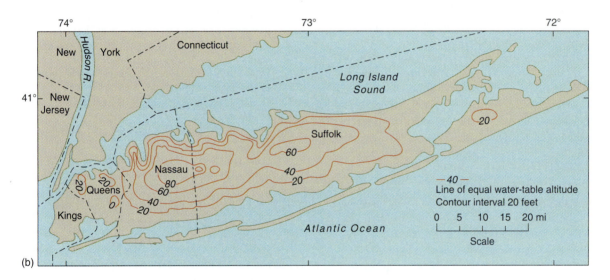

(b)

Map for problem 5.

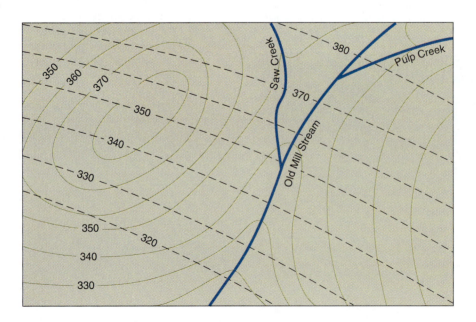

Figure 11.10a

Land subsidence in the San Joaquin Valley, California, 1926–1970. Most of the subsidence resulted from groundwater withdrawal, but some is due to petroleum removal as well. *Source: R. L. Ireland, 1984, p. 4, U.S. Geological Survey.*

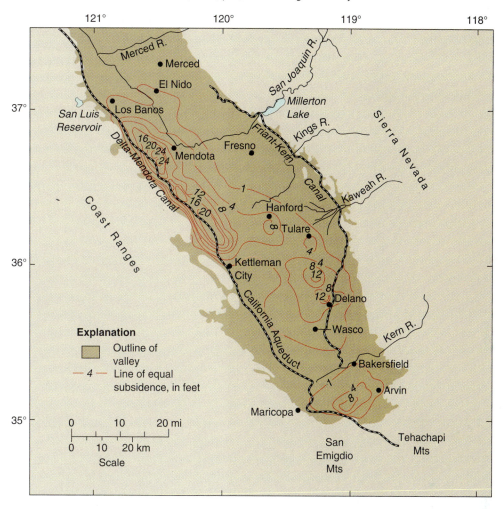

g. Suppose you were going to purchase one-quarter of this map area on the basis of groundwater supply. Which would you choose: NW, NE, SE, or SW? Why?

6. What volume of water would flow through a valley filled with porous, permeable quartz sand 100 ft thick and 1 mi wide, where the permeability is 500 ft/day and the potentiometric gradient is 5 ft/mi?

7. In Figure 11.5, describe the size, shape, and depth of the cone of depression since 1880.

8. With reference to Figure 11.10a, land subsidence in the San Joaquin Valley,
   a. What has been the minimum average rate of subsidence per year in the valley?
   b. Why do you think the areas of maximum subsidence are located near the coastline?
   c. Describe the likely effect of land subsidence in the San Joaquin Valley on the regime of rivers draining west-to-east from the Coast Ranges. How might erosion and sediment transport be affected?

d. The San Joaquin Valley produces a large percentage of the fruits and vegetables for the American people, a circumstance made possible by extensive irrigation. Restricting farmers' use of subsurface water would drive many of them into bankruptcy, with associated harmful effects to agriculture-based industries. What should the state of California do to balance the competing needs of the agricultural community, city dwellers, and factory owners whose buildings are being destroyed by the ground subsidence?

Further Reading/References

Heath, R. C., 1983. *Basic Ground-water Hydrology.* U.S. Geological Survey Water-Supply Paper 2220, 84 pp.

Heath, R. C., 1989. *Ground-water Regions of the United States.* U.S. Geological Survey Water-Supply Paper 2242, 78 pp.

Higgins, C. G., and Coates, D. R. (eds.), 1990. *Groundwater Geomorphology.* Geological Society of America Special Paper 252, 368 pp.

Figure 11.10b

Magnitude of subsidence at a site 10 miles southwest of Mendota, San Joaquin Valley, California. Power pole shows position of land surface in 1925, 1955, and 1977. Land surface was lowered about 30 feet during that period.

Ireland, R. L., Poland, J. F., and Riley, F. S., 1984. *Land Subsidence in the San Joaquin Valley, California, as of 1980.* U.S. Geological Survey Professional Paper 437-I, 93 pp.

Newton, J. G., 1987. *Development of Sinkholes Resulting from Man's Activities in the Eastern United States.* U.S. Geological Survey Circular 968, 54 pp.

Palmer, A. N., 1991. "Origin and morphology of limestone caves." *Geological Society of America Bulletin,* v. 103, pp. 1–21.

Postel, Sandra, 1992. *Last Oasis: Facing Water Scarcity.* New York, W. W. Norton, 239 pp.

WATER POLLUTION

The volume of water on the earth is limited; existing supplies must be carefully managed in terms of both conservation and cleanliness. Both considerations are necessary if the world is to avoid a catastrophe of unimaginable proportions. Conservation can be achieved through public education, but keeping our water supplies clean is a much more difficult problem. Unless more effort is expended on keeping our water pure, we might someday need to boil, filter, and decontaminate the water in our homes before we use it. Such a prospect sounds like science fiction, but it will become all too real if the present trend toward increasing water pollution continues.

Pollutants or contaminants can consist of any of the following, each of which creates different problems for treatment:

1. Microorganisms, including pathogenic viruses and bacteria

2. Organic matter, primarily from domestic urban sewage and from rural septic tanks

3. Chemical wastes from chemical plants and industrial operations such as mining and petroleum exploration; from leaks in pipelines and storage tanks; from agricultural operations, such as runoff from animal feedlots and fields treated with pesticides and fertilizers; and from leaching of not-so-sanitary landfills

4. Nuclear wastes generated by power plants, weapons manufacturing plants, laboratories, and medical research facilities.

WATER-QUALITY STANDARDS

Requirements for water purity depend on the use to which the water is to be put. Drinking requires the highest purity; irrigation has lower standards, at least for many substances. The federal government has developed criteria for all categories of water quality, including bacterial content, physical characteristics, and chemical constituents (Table 12.1). Normally, problems involving bacterial content or physical characteristics can be alleviated. Removing or neutralizing undesirable chemical contaminants, in contrast, is often both difficult and expensive. The presence of chemical impurities often imposes major limitations to water utilization. Removal of many organic pollutants poses problems that are as yet unsolved.

In addition to difficulties involving measurable contaminants, we face the rarely mentioned problem of substances present in amounts too small to detect with current equipment. For example, for some pollutants the detection limit may be in parts per million but the substance may be harmful in amounts as low as parts per billion. A related, rarely discussed problem is that we know so little about the effects of many substances on humans that we have set no drinking-water standards for those substances. Without a perfect knowledge of human biochemistry, which can never be achieved, we can never be certain which substances in which amounts endanger human health. Unless we completely dismantle our industrial civilization and return to living in caves, however, we will always be poisoning ourselves to a greater or lesser extent. We cannot

TABLE 12.1

Maximum Allowable Concentrations of Dissolved Inorganic Substances in Drinking Water According to the Environmental Protection Agency

Contaminant	Maximum Allowable Level (ppm)
Antimony	0.01
Arsenic	0.05
Barium	1.00
Boron*	1.00
Cadmium	0.01
Chloride*	250.00
Chromium	0.05
Copper*	1.00
Fluoride	1.4–2.4
Hydrogen sulfide*	0.05
Iron*	0.30
Lead	0.05
Manganese*	0.05
Mercury	0.002
Nitrate (as N)	10.00
Selenium	0.01
Silver	0.05
Sulfate*	250.00
Zinc*	5.00
Total Dissolved Solids*	500.00

*Values for substances with asterisks are the EPA-reasonable goals for drinking water but are not federally enforceable. Many of the elements listed are essential for human nutrition in small amounts. Standards also exist for bacteria and for many industrial organic compounds.

solve the pollution problem completely; instead, we must do the best we can with our current tools and understanding.

RESIDENCE TIME

When we find that an area or aquifer is polluted, we must answer three questions:

1. What is the source of the pollutant? Sources can be of two types: point sources such as a septic tank, an oil spill, or an industrial waste outlet; or nonpoint sources such as farmland runoff or strip-mine drainage. Point sources are much easier to find and remedy.

2. What is the nature of the pollutant? Is it a single element such as arsenic, cadmium, or lead? A complex organic compound such as PCBs (polychlorinated biphenyls) or vinyl chloride? A mineral group such as asbestos? A radioactive by-product from an industrial or military project? Each of these materials poses different treatment problems.

3. What are the migration pattern and expected lifetime of the pollutant? Pollutants can be solids, liquids, or gases. Many are destroyed quickly in the natural environment, perhaps by interaction with bacteria. Radioactive materials can self-destruct to an acceptably low level within a few days or can linger for millions of years. But the lifetimes of many industrial and chemical pollutants are unknown and, for safety reasons, must be assumed to be very long.

When we evaluate a polluted aquifer, we often express the duration of the pollutant as its residence time, defined as

$$R = \frac{C}{F}$$

where R is the residence time, C the capacity of the reservoir, and F is the rate of inflow and outflow of the compound. Because inflow and outflow rates can change, the residence time is strictly accurate only at the time the measurements are made. Note that this definition of residence time does not consider that the pollutant might decompose or break down into other chemicals before it outflows. Note also that this concept of residence time does not include the fact that the pollutants are likely to be absorbed on clays and organic matter in the water. As the water becomes less contaminated, the contaminants will be gradually released from these host materials, greatly increasing their residence time in the water.

GROUNDWATER

The widespread use of chemical products, coupled with the disposal of large volumes of waste materials, creates the potential for extensive groundwater contamination. Some of the most prominant areas of contamination, such as Love Canal in upper New York state, have attracted public notice. But for every one of these there are a large number of smaller contamination problems that have until very recently gone unnoticed. All students of groundwater pollution problems agree that episodes such as Love Canal are only the tip of the iceberg. New instances of groundwater contamination are being identified continuously in urban, industrial, and agricultural settings. Many of these pollution problems have existed for some time but are only now being identified, thanks to our developing analytical capabilities and increasing concern about the effects of impure water on human health and the environment.

Mining activities in the western United States, for example, started during the middle 1800s, when Americans worried little about the environment and knew almost nothing about the effects of chemical pollutants on human health. As a result, mining debris was not dealt with properly. Now, however, that same debris is considered hazardous to humans and to other animal and plant life. In some areas of the American

West, hazardous levels of heavy elements such as lead, zinc, mercury, or chromium occur in surface or groundwater supplies. Currently applicable laws state that the current owner of the mining property is legally and financially responsible for damage to the water supply and related human health, regardless of whether the mine is still active or whether the present owner had even been born when the pollution occurred.

The pollution problems left over from the last century, however, pale in comparison with those being created by the immense volumes of toxic inorganic and synthetic organic materials produced by modern industries, both in the United States and overseas. Some areas of eastern Europe have already been made uninhabitable by pollution of the past 50 years. Many industrially produced chemicals are quite stable in groundwater and pose a serious threat to human health. Current estimates suggest that 0.5–2% of the groundwater in the conterminous United States is contaminated from point sources. This estimate based on localized sources does not include contamination from nonpoint sources such as agricultural runoff. Certain areas, especially those with a high population density, may have much higher percentages of contaminated groundwater. The EPA has a growing list of more than 400 contaminated water sites.

GROUNDWATER CONTAMINANTS

The toxic materials that enter groundwater come from many sources (Figure 12.1), including septic systems, runoff from agricultural fields and animal feedlots, landfills, accidental leaks and spills, mining debris, ruptures in underground storage tanks such as those beneath gasoline stations, decomposing bodies in graveyards, disposable diapers, cheesy pizza sludge, and underground injection of hazardous waste.

Given their wide variety of sources, it is not surprising that the types of contaminants are extremely varied. Some are simple inorganic ions such as nitrate from fertilizer and feedlot wastes, chloride from deicing salt and saltwater intrusion, and heavy metal ions from plating works and many other industrial processes. Other contaminants are more complex synthetic organic compounds that result from industrial and manufacturing processes and the use of pesticides and household cleaners. In many instances we know nothing about the chemical stabilities and lifetimes of these compounds, or their effects on human biochemistry.

The magnitude of any pollution problem depends on the size of the area affected; the amount of pollutant involved; the solubility, toxicity, and density of the pollutant and its persistence in the environment; the mineral composition and permeability of the soils and rocks through which the pollutant moves; and the potential effect of the pollutant on groundwater use. For example, if groundwater contaminants exceed the federal standards for drinking water, then the water is considered hazardous to drink. However, these standards include only a limited number of chemicals and do not protect humans and the environment against all possible contaminants. The long-term effects of even small amounts of many contaminants are unknown.

Figure 12.1

Schematic representation of contaminant plumes possibly associated with various types of waste disposal. *National Research Council, 1984, page 7*, Groundwater Contamination.

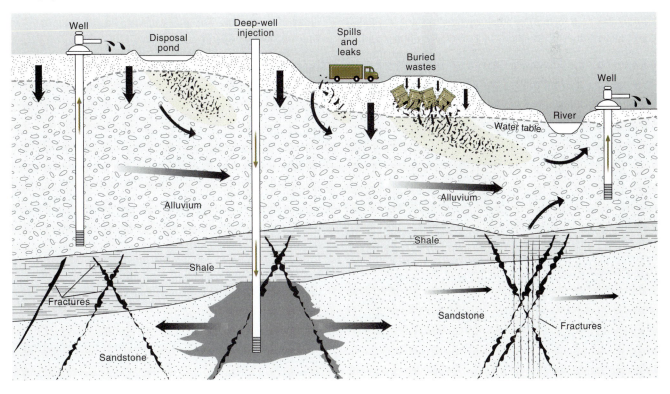

Given these complexities, we can prevent significant groundwater pollution only by selecting waste-disposal sites in such a way that

1. An adequate thickness of unpolluted sediment containing clay and/or organic material is present both above and below the waste. Both clay and organic matter absorb trace metals and many complex organic contaminants, preventing these substances from entering the groundwater.

2. The disposal areas are as close as possible to places of natural groundwater discharge so that any contaminant entering the groundwater will be washed out quickly. Of course, this contaminated surface discharge must not be permitted to flow for long distances overland or infiltrate downward into the subsurface.

CONTAMINANT MOVEMENT

Most aquifers are not homogeneous, making times for contaminant migration through subsurface materials difficult to calculate accurately. Complications arise from variable geologic conditions (Figure 12.2) and from the many types of contaminants. Layered beds and lenses of less-permeable rock within an aquifer can cause fingering and separation of a contaminant plume, and the grains that form the aquifer cause the contaminant to disperse and be diluted within the aquifer (Figure 12.3). Contaminant-absorbing materials such as clay minerals may be irregularly distributed within the host rock. Contaminants from different sources may react chemically within an aquifer to create new substances, or microbes may interact with and reduce the amounts of some

Figure 12.2

Possible consequences of subsurface injection at a site not having the necessary hydrogeologic characteristics. *Source: U.S. Geological Survey Water-Supply Paper 2281, 1986.*

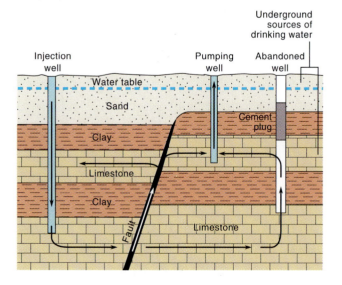

Figure 12.3

(a) Relationship between the true flow path of a fluid in a clastic rock and the mean flow path. (b) Dispersion of the flow in the rock, illustrating the way an initially narrow band of pollutant spreads throughout the rock layer. *R. Allen Freeze/John A. Cherry, Groundwater, © 1979, pp. 70, 384. Reprinted by permission of Prentice Hall, Englewood Cliffs, New Jersey.*

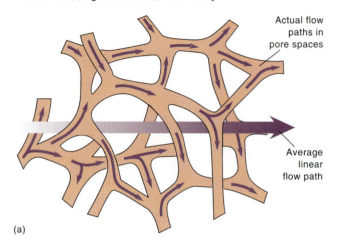

(a)

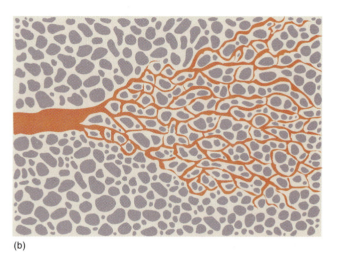

(b)

pollutants. Some liquid contaminants, such as sea water, and some synthetic organic compounds are heavier than fresh water so that they sink and concentrate along the base of the aquifer. Others that are lighter, such as gasoline, float and concentrate toward the top. Unfortunately, variations in both geologic conditions and the contaminants occur below the surface where we cannot observe them directly. Groundwater becomes very difficult or even impossible to purify once it has been contaminated. And even if we have the technology to decontaminate an aquifer, the financial cost may well be so high that we must write off the aquifer as permanently unusable. This has already occurred in some localities.

RADON

Radon is a naturally occurring colorless, odorless, and tasteless *radioactive* gas that forms during the natural disintegration of uranium atoms. Minor amounts of uranium are

Figure 12.4

Major radon entry routes into houses. *Source: Environmental Protection Agency.*

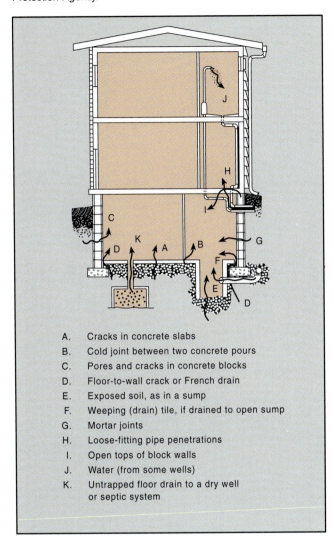

A. Cracks in concrete slabs
B. Cold joint between two concrete pours
C. Pores and cracks in concrete blocks
D. Floor-to-wall crack or French drain
E. Exposed soil, as in a sump
F. Weeping (drain) tile, if drained to open sump
G. Mortar joints
H. Loose-fitting pipe penetrations
I. Open tops of block walls
J. Water (from some wells)
K. Untrapped floor drain to a dry well
    or septic system

present in all rocks and soils and, therefore, in many groundwaters. In addition, radon leaking upward into houses and being trapped there may be a health hazard (Figure 12.4). The amount of radon that most people are exposed to is very small and the degree of danger from exposure to these low levels is uncertain. The federal government believes the danger is significant for a large number of Americans. Many scientists doubt this assessment.

Problems

1. What would be the advantages and disadvantages of locating a sanitary landfill on a river floodplain?

2. Figure 12.5 is a map of an area underlain by a moderately sorted fine-grained sandstone that contains occasional lenses of clay. Based on the static water levels in wells not shown on the map, the locations of natural springs, the

Figure 12.5

Map of an area two miles to the east of Milleville showing surface drainage, elevation of the water table at 24 locations, and several man-made features.

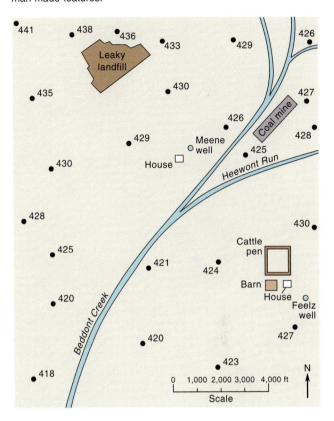

geologic map of the area, and scattered data from previous geologic investigations, the elevation of the water table is known at 24 locations in the map area. It ranges from 418 ft in the southwest corner of the map to a maximum of 441 ft in the northwest corner. At most of these 24 sites, the water table is 10–12 ft below the ground surface. Based on these data,

a. Draw a contour map of the elevation of the water table in the map area using a contour interval of two feet.
b. What is the hydraulic gradient between the landfill and Mr. Meene's house?
c. In which direction does Beddont Creek flow? How do you know?

Also located on the map are four built features:
*   A cattle ranch owned by Mr. Feelz, a retired geologist, that has a pen, barn, house, and water well.
*   A small homestead owned by Mr. Meene, who draws water from both his water well and Beddont Creek, depending on the amount of seasonal rainfall.
*   An abandoned strip mine from which coal was excavated until 1980.
*   A leaking landfill that serves Milleville, a small town just west of the map area. The landfill is operated by a locally owned company and is excavated to a depth of about eight feet below the ground surface.

Figure 12.6a

Location of study areas and sampling sites. Symbols indicate levels of radon-222 concentrations. *Source: U.S. Geological Society.*

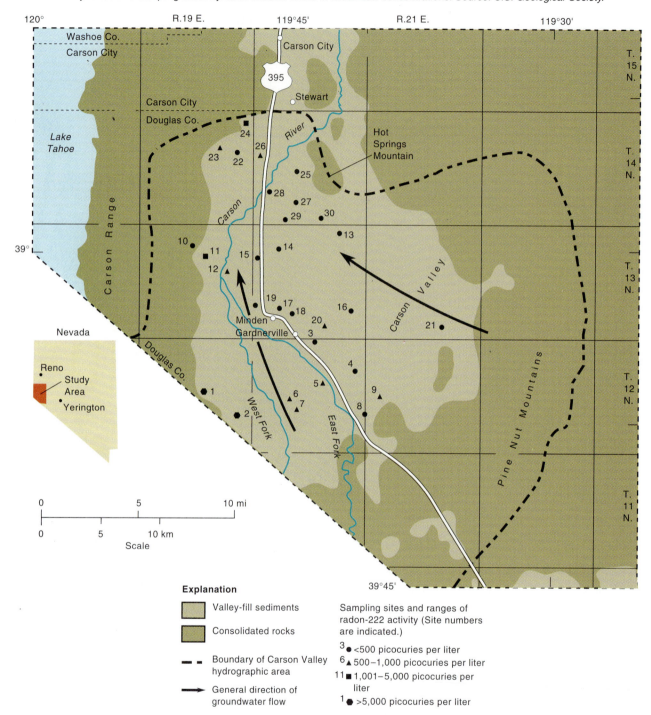

Explanation

Valley-fill sediments

Consolidated rocks

- - - Boundary of Carson Valley hydrographic area

→ General direction of groundwater flow

Sampling sites and ranges of radon-222 activity (Site numbers are indicated.)

3 ● <500 picocuries per liter

6 ▲ 500–1,000 picocuries per liter

11 ■ 1,001–5,000 picocuries per liter

1 ⬣ >5,000 picocuries per liter

As a local environmental specialist in good standing, you have been asked by the Milleville town council to answer the following questions.

d. Is there any danger that refuse in the leaking landfill will contaminate Mr. Meene's or Mr. Feelz' groundwater supply? Why or why not?

e. If you believe that either person's well might become contaminated from the landfill, what would you recommend? Must the landfill be abandoned? Should a new one be located in the map area? If so, where would you recommend it be located and why?

f. Is Mr. Meene's water supply in danger of contamination from Mr. Feelz' cattle pen? Explain.

g. Might the abandoned coal mine contaminate either person's water supply? If so, which person is more likely to be harmed?

Figure 12.6b

Cross-section of hydrogeologic relations in Carson Valley. Upper part of basin-fill section is Quaternary age and lower part is Tertiary age. View looking north. *Source: U.S. Geological Survey.*

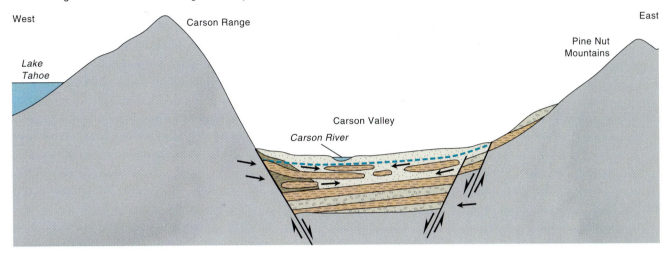

Explanation

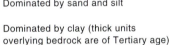 Fluviatile and lacustrine deposits
(groundwater can be confined or unconfined)

Dominated by cobbles and gravel

Dominated by sand and silt

Dominated by clay (thick units
overlying bedrock are of Tertiary age)

Alluvial-fan deposits (poorly sorted:
groundwater generally confined)

Bedrock

Major fault (arrows show relative
direction of movement)

Groundwater flow path

- - - Water table

3. The major source of domestic drinking water near Carson City, the capital of Nevada, is shallow groundwater from neighboring Carson Valley. The U.S. Geological Survey surveyed radon contents in the groundwater at the request of the state government. The areal geography, sampling sites, and survey results are shown in Figure 12.6a. A cross-section of the local topography and valley fill are shown in Figure 12.6b, and the generalized geology of the area in Figure 12.6c. Use these maps to answer the following questions.

   a. The drainage basin for the water ("hydrographic area") is larger east of the Carson River than to the west. Would you have expected this to be true based on the appearance of Figure 12.6a? Explain.

   b. How many potential aquifers does the cross-section show?

   c. Which of these potential aquifers do you think is the major water source for private homes? Why?

   d. What part of Carson Valley has the highest radon levels? Explain this result in terms of the geology of the area.

Further Reading/References

Eisinger, Joseph, 1996. "Sweet poison." *Natural History,* v. 106, no. 7, p. 48–53.

Hamilton, P. A., and Shedlock, R. J., 1992. "Are fertilizers and pesticides in the groundwater?" U.S. Geological Survey Circular 1080, 15 pp.

Moore, J. W., and Ramamoorthy, S., 1984. *Heavy Metals in Natural Waters.* New York, Springer-Verlag, 268 pp.

Patrick, R., Ford, E., and Quarles, J., 1987. *Groundwater Contamination in the United States,* 2nd ed., Philadelphia, University of Pennsylvania Press, 513 pp.

Smith, J. A., Witkowski, P. J., and Fusillo, T. V., 1988. *Manmade Organic Compounds in the Surface Waters of the United States—A Review of Current Understanding.* U.S. Geological Survey Circular 1007, 92 pp.

U.S. Environmental Protection Agency, 1991. *Fact Sheet, Drinking Water Regulations Under the Safe Drinking Water Act,* June, Washington, D.C.

Yanggen, D. A., and Webendorfer, B., 1991. "Groundwater protection through local land-use controls." Wisconsin Geological and Natural History Survey Special Report 1, 48 pp.

Figure 12.6c

Generalized geology of the study area and adjacent areas. *Source: U.S. Geological Survey.*

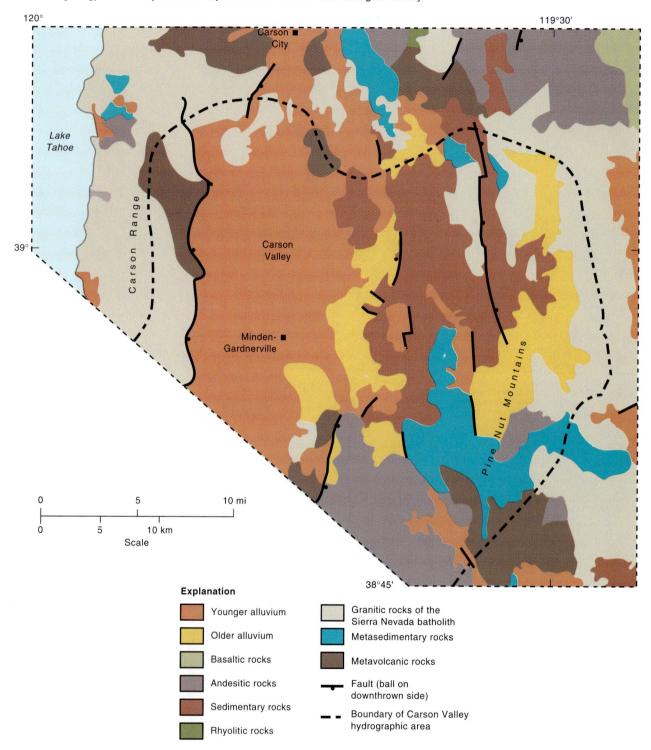

Explanation

Younger alluvium

Older alluvium

Basaltic rocks

Andesitic rocks

Sedimentary rocks

Rhyolitic rocks

Granitic rocks of the
Sierra Nevada batholith

Metasedimentary rocks

Metavolcanic rocks

Fault (ball on
downthrown side)

Boundary of Carson Valley
hydrographic area

E X E R C I S E 13

SOLID WASTE DISPOSAL

People produce garbage, a staggering array of waste that includes paper, wood, plastic, metal, rubber, and a host of other manufactured products (Figure 13.1). The more "up to date" a society is, the more waste it produces. Even worse, as a society becomes more "advanced," less of the waste is composed of things that decompose rapidly (Figure 13.2). Most of our refuse is not *biodegradable* within a reasonable length of time when left on the ground. It hardly decomposes at all if buried and removed from contact with air and water. Rubber tires and plastics last indefinitely, unlike chicken bones and cloth sandals. Americans generate a monstrous total of 550,000 tons of solid waste every day, an average of 4.5 pounds per person. Nearly all of it will last hundreds to thousands of years.

WHAT DO WE DO WITH IT?

There are three ways to dispose of this stuff: (1) burying it in *sanitary landfills;* (2) burning it in huge incinerators; and (3) recycling it to slow the need for new production. At present, about 60% of our waste is buried, 20% is burned, and 15% is recycled. The other 5% is simply tossed somewhere by uncaring individuals.

There are now 75,000 industrial landfills, 5,800 operating municipal landfills, and about 40,000 closed or abandoned municipal landfills. Needless to say, landfills, which are essentially giant holes in the ground, fill up rapidly around large urban areas. A city of one million people produces four and a half million pounds of waste each day. Many cities send their waste to relatively uninhabited areas in other parts of the country, at great expense (Figure 13.3). A growing part of city taxes goes toward garbage disposal. It commonly costs large cities $100 per ton to get rid of their waste.

Surveys indicate that American households deposit about 350,000 gallons of legally *hazardous liquids* each year in landfills. In addition, most municipal garbage contains some hazardous solid materials such as paint solvents, batteries, and insecticides. These materials can release heavy elements and poisonous synthetic organic compounds. For these reasons, modern municipal landfills have a heavy plastic or rubbery liner at their base. The liners are designed to prevent rainwater and liquids that descend through the pile from dripping into underlying soil and groundwater. Unfortunately, a large percentage of operating landfills are not leakproof, either because they were poorly designed or because they have developed leaks over time.

As organic materials decompose in a landfill, they produce carbon dioxide and methane. Often the amount of methane is large enough to be captured by pipes inserted into the landfill and used as a fuel. The relative amounts of carbon dioxide and methane depend on the amount of air (oxygen) that penetrates the landfill. The more oxygen, the more carbon dioxide and the less methane.

Figure 13.1

Waste from a civilized society.

Figure 13.2

Percentage by weight of materials in municipal solid waste. *Source: Environmental Protection Agency.*

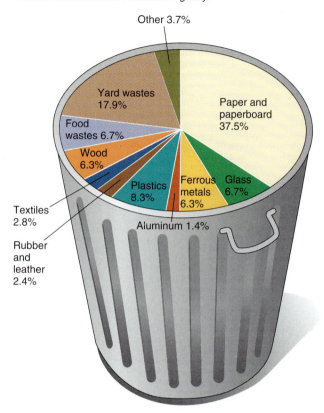

Incineration reduces the amount of municipal waste by 90% and has the added benefit that the heat produced can be used to generate electricity. However, if the emissions from the smokestacks are not controlled, acids, dangerous chemicals, and toxic heavy metals can be released into the atmosphere. In addition, incinerators in the United States now produce 5 million tons of ash every year. This ash has a higher percentage of toxic elements than the material that entered the incinerator.

Recycling is growing in popularity in the United States. About 35–40% of our steel and aluminum is recycled, as is 25% of used paper. Unfortunately, recycling is often not cost-effective. Many cities that established recycling programs have abandoned them for this reason. But as landfill space grows more expensive and the cost of environmentally acceptable incinerators climbs, the cost of recycling will become more competitive nationwide.

TOXIC WASTE SITES

There are thousands of waste dumps and landfills that store hazardous wastes. Many of them store wastes in an unsafe manner, in open lots. Steel drums rust in the rain and waste leaks into local water supplies. Many cases of this pollution have been documented. Medical studies have shown that living near a hazardous waste site increases your risk of cancer and of having children with birth defects. About 16% of Americans live within four miles of a *toxic waste site*. The Environmental Protection Agency estimates that America has more than 400,000 problem waste sites with chemical storage tanks that leak organic chemicals and pesticides into soil and water.

Because of this environmental disaster, Congress passed a law, popularly known as *Superfund,* to clean up the 1,300 worst of these blights. About 100 new superfund sites are added each year, far faster than cleanup is removing old sites from the list. The 1996 budget for Superfund was $6.1 billion. Money devoted to Superfund is certain to increase significantly over the coming years.

RADIOACTIVE WASTE DISPOSAL

Radioactive waste is dangerous to living creatures for periods ranging from hundreds to millions of years. About one third of the radioactive waste volume is classed as low-level radwaste, meaning that it will be dangerous for 300–500 years. It comes not from nuclear reactors but from hospitals, university and industrial research laboratories, and military

Figure 13.3

Interstate traffic in garbage, excluding New York and New Jersey. New York's waste is shipped as far away as New Mexico. *Source: National Solid Waste Management Association.*

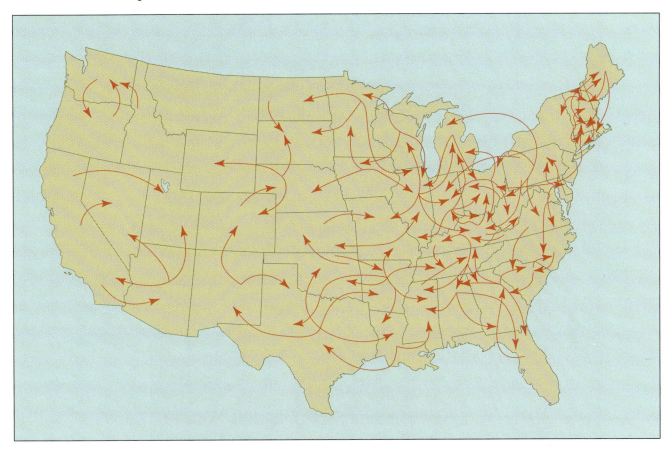

facilities. The materials include fabric, glass, metal, plastic, paper, wood, and animal remains.

High-level radioactive waste will be dangerous for thousands to millions of years, far longer than any human civilization has existed. It forms two thirds of the radwaste volume but contains 94% of the radioactivity. It is produced entirely by nuclear reactors, mostly by the military but also by industrial power plants.

DISPOSAL SITES

For both low-level and high-level radwaste, the main concern is to keep the radioisotopes from entering the water supply, as has occurred at some disposal sites (Figure 13.4). For this reason , current scientific opinion favors disposal of low-level waste in excavations above the groundwater table in arid or semiarid areas, such as those in the western and southwestern United States. These disposal sites offer several advantages: the water table is hundreds of feet below the present ground surface and is very unlikely to rise significantly over the next 300–500 years; little water percolates downward in such dry regions; cavities tens of feet deep can be excavated easily; monitoring of excavated storage cavities is simple; and backfill from the cavities can be removed easily should problems develop tens or hundreds

of years from now. In addition, the natural advantages of these disposal sites could be further improved by barriers engineered to prevent water infiltration.

High-level radwaste poses more difficult problems because of the very long periods during which it is radioactive. The waste is too dangerous to be stored above ground for extended periods and disposal methods such as rocketing it into space, dropping it into cracks in glaciers, or depositing it on the ocean floor are either too dangerous or too expensive. At present, the favored disposal method for high-level radwaste is underground storage. Current plans call for filling numerous cannisters with hot, highly radioactive material, putting them deep underground, and keeping them there in isolation for at least 10,000 years.

But how will such prolonged exposure to heat and radiation affect the rocks enclosing the repository? Can the cannisters be kept from water indefinitely? Will they corrode? How fast will the radioisotopes dissolve and possibly move out of the containment area? Currently, we have no reliable answers to such questions. It is, perhaps, sobering to recall that no human civilization has ever lasted for anything close to 10,000 years, most disappearing within a few hundred years.

Figure 13.4

Cross-section showing the vertical distribution of tritium concentrations in groundwater near burial site Plot M near Chicago, Illinois, in 1981 and 1983. The low-level radwaste was buried from 1943 to 1949. Background tritium concentration is 0.2 nCi/L. Vertical lines are wells from which data were obtained.

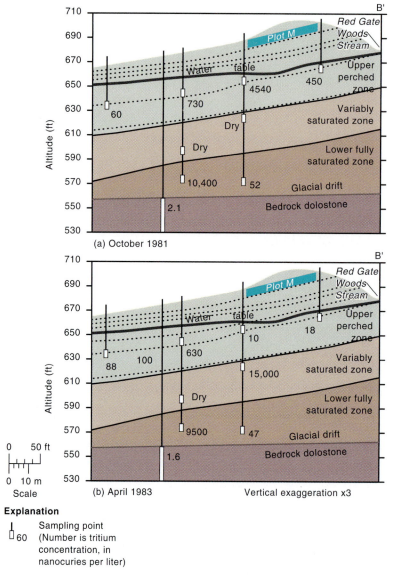

Explanation

Sampling point
(Number is tritium
concentration, in
nanocuries per liter)

• • • • • Sand layer

Burial sites for high-level radwaste have clear geologic requirements. The rock at the appropriate depth must be strong enough to maintain an opening, should have low permeability and few fractures through which groundwater might move, and should not be in an area prone to earthquakes, volcanic activity, or severe erosion. Rocks best suited to radwaste storage should also have high heat conductivity to help keep temperatures low in the containment area. The rock types that best satisfy all these criteria are bedded salt or salt domes, granite, basalt, argillaceous rocks, and tuffaceous rocks.

The scientific problems concerning radwaste disposal are serious, but even more difficult are the political problems. Most people are unwilling to live near a disposal site.

Americans fear radioactivity so strongly that no assurances by scientific or governmental offices seem adequate. Almost everyone agrees that radwaste disposal is necessary but few people want a disposal site nearby. Part of the fear stems from the invisibility of radiation and the natural fear of unseen things, and part from distrust of governmental pronouncements. The federal government has often underestimated pollution dangers to the public (as well as exaggerating some others), and people have become understandably wary. But radioactive waste must be placed somewhere; allowing it to remain above ground is even more dangerous than storing it below ground. The Environmental Protection Agency has an important but difficult education job to do on this issue.

Problems

1. As an average American, you generate 4.5 pounds of waste per day. The average Japanese, German, or Swede generates less than half this amount. Why do you believe Americans are so wasteful? List the benefits to American society of decreasing the amount of solid waste we produce.

2. What are the benefits and drawbacks of incineration as a method of disposing of solid waste?

3. The five states that lead the nation in the production of hazardous waste are New Jersey, Pennsylvania, California, New York, and Michigan. What factors might these states have in common that has resulted in their infamous leadership role?

4. How can both landfills and incinerators serve as sources of energy?

5. Examine a geologic map of the area where you live. Locate the place where your city's landfill is located. How did the city choose this site? Do you believe this location was a good choice? Explain why. Are there other places within a few miles of the city that would also be satisfactory? Why?

6. Figure 13.4 shows the result of poor disposal practices in the 1940s at a site near Chicago, Illinois.

a. What is the approximate thickness of the glacial drift?

b. What do you think controls the thickness of the "variably saturated zone"?

c. Tritium is a radioactive isotope of hydrogen, H3. Contour the tritium values at a logarithmic contour interval of powers of ten (10, 100, 1000, etc.).

d. The contours of tritium values do not form a concentric hemispherical pattern below the disposal site, as might be anticipated. They form elongated fingers. What does this pattern indicate about the internal structure of the glacial drift?

e. The contours appear compressed at the boundary between the drift and the underlying dolostone. What might explain this pattern?

f. The dolostone contains both horizontal and vertical joints. What effect might these have on tritium migration?

g. Presumably the people who live near Plot M are not overjoyed about the high tritium levels in their groundwater supply. As county supervisor, you receive many complaints about the real or imagined dangers of imbibing tritium in amounts thousands of times higher than the background amount. How do you deal with these complaints?

Figure 13.5

Surface sediments near Fort Lupton, Colorado. *From John E. Costa and Victor R. Baker,* Surficial Geology. *Copyright © 1981 John Wiley and Sons, Inc., New York, NY. Reprinted by permission of John Wiley & Sons, Inc.*

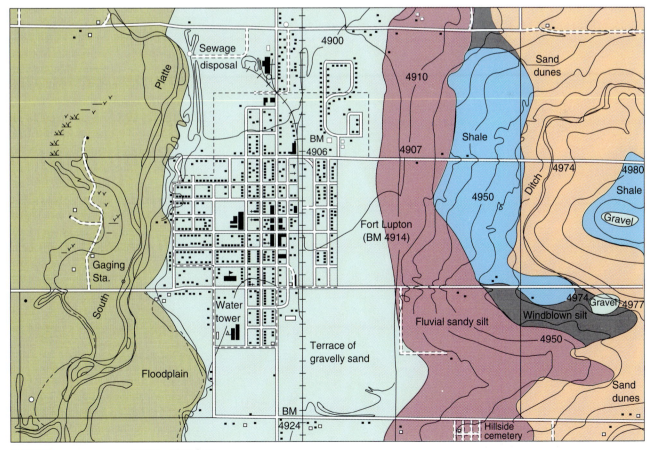

1 : 24,000   1 inch = 2000 feet

10 ft contour interval

h. As a resident of the area, should you consider relocating? In which direction and how far should you move? (You work in Chicago about 20 miles northeast of your home.)

7. Figure 13.5 is a topographic map of Ft. Lupton, Colorado, on which is overlain some geologic information.
   a. Locate suitable sites for sanitary landfills and explain your choices.
   b. The population of Ft. Lupton is 5,200 and each resident generates seven pounds of solid waste each day. How many pounds of waste accumulate each year?
   c. Experience has revealed that, on average, 40 pounds of Ft. Lupton's waste can be compacted into one cubic foot in the landfill. What volume accumulates in the landfill each day? Each month? Each year?
   d. The company that has the contract to dispose of the city's waste covers each day's deposit with dirt. The daily volume is four parts solid waste and one part fill. What is the amount of dirt used each day? How much each year? Locate a suitable source of dirt for the landfill.
   e. If the city excavates a pit 300 feet by 300 feet and 15 feet deep, in how many years will the landfill become a landfull, assuming no population growth?

8. You are interested in the possible effects of natural radiation on the frequency of birth defects in children.
   a. Design a program of data collection to study this question.
   b. Suppose your investigation reveals that pregnant women in certain parts of your state are twice as likely to give birth to children with birth defects as women in other parts of the state. What do you think the government should do with this information? Do your results have any financial implications for the state's population?
   c. Suppose a nationwide study of the type you have made indicates that stillbirths and birth defects are twice as frequent in Colorado as in neighboring Nebraska. Should the federal government develop a policy to deal with this "unfair" variation among states? If so, what?

Further Reading/References

Grove, Noel, 1994. "Recycling." *National Geographic,* July, pp. 92–115.

Hasan, Syed E., 1996. *Geology and Hazardous Waste Management.* New York, Prentice-Hall, 387 pp.

Krauskopf, Konrad B., 1988. *Radioactive Waste Disposal and Geology.* New York, Chapman and Hall, 145 pp.

La Sala, A. M., Jr., et al., 1985. *Radioactive Waste. Issues and Answers.* Arvada, Colorado, American Institute of Professional Geologists, 27 pp.

O'Leary, Phillip R., Walsh, Patrick W., and Ham, Robert K., 1988. "Managing Solid Waste." *Scientific American,* Dec., pp. 36–42.

Rathje, William L., 1991. "Once and Future Landfills." *National Geographic,* May, pp. 116–34.

Whipple, Chris G., 1996. "Can nuclear waste be stored safely at Yucca Mountain?" *Scientific American,* June, pp. 56–64.

14

COASTAL EROSION

The coastal zone is an area of variable width that includes land and sea areas close enough to the shoreline to be affected by nearshore processes (Figure 14.1). In some areas of the United States, the coastal zone is tens of miles wide. Processes that affect this zone include waves, currents, hurricanes (Figure 14.2), tides, tsunamis, and floods at the mouths of large rivers such as the Mississippi. More than 25% of the American population lives in areas close enough to the shoreline to be affected by at least one of these processes, and the percentage is increasing yearly. This high percentage of nearshore dwellers stems in part from the locations of the seaports where most of the nation's immigrants landed between 1860 and 1920, and in part from the aesthetic value of many coastal areas. Coastal communities account for more than half the residential construction at present. By 2010 the coastal population of the U.S. is expected to reach 127 million.

SAND MOVEMENT

Human activities continually interfere with natural coastal processes, causing pollution and ecologic changes; conversely, the same natural processes interfere with people's enjoyment of the seashore. Foremost among the natural processes is sand movement and the resultant destruction of beaches and barrier sand bars that have both commercial and aesthetic value. Beach sand is lost permanently through transportation into deeper water, beyond the reach of shoreward wave and current action, through transportation by onshore winds to inland dunes, and through pulverization into particles too small to remain on the beach.

The bulk of sediment movement along coastlines results from waves and the longshore currents they spawn. Wavelength depends on wind strength and duration, the distance over which the wind blows (the fetch), and the configuration of the shore bottom. The water molecules within a wave move in a circular pattern, with a net forward motion toward the shore. When the water depth becomes less than about one-half the wavelength, the circles start to scrape on the sea bottom (Figure 14.3). This contact slows the lower part of the water mass, while the upper part rushes forward with an accompanying increase in wave height. Because the energy of the onrushing wave increases as the square of wave height, high waves have very large erosive capabilities. During storms and winter months wave action moves much more sediment offshore than is replaced during nonstorm and summer months.

If the seafloor surface in the nearshore area remained perfectly smooth as water depth increased and if the wind blew exactly perpendicular to a straight shoreline, then all the energy of the wind-generated waves would be expended normal to the beach. In the real world, however, such a situation never occurs; instead, some wave energy is focused parallel to the beach and generates currents, called *longshore currents,* that move sediment parallel to the shore (Figure 14.4). Measurements on the east and west coasts of the United States give a range of transport volumes from 10^4 to 10^6 m^3/yr. Most sand movement occurs in the breaker zone because of the high kinetic energies there. In California, for example, 80% of the longshore drift occurs in water less than 2 m deep. In general, high waves transport sand in the breaker

Figure 14.1a

Coastal erosion in the United States. All 30 coastal states are experiencing erosion along their coastlines. *Source: U.S. Geological Survey.*

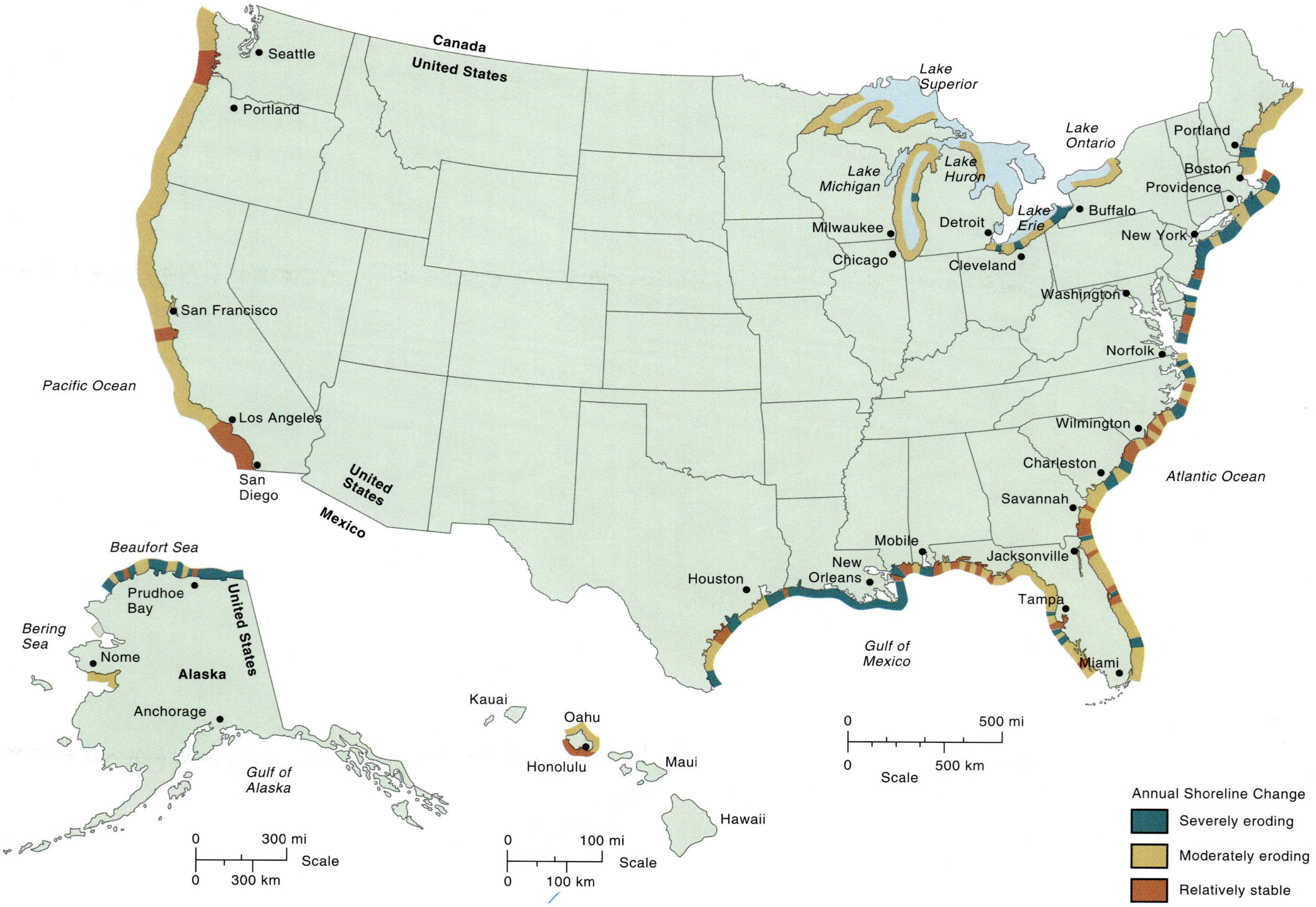

Annual Shoreline Change

Severely eroding

Moderately eroding

Relatively stable

Figure 14.1b

Student housing.

zone; low waves move sand mostly in a zigzag fashion along the beach face.

Along open coasts the currents caused by the twice-daily rise and fall of tides are rarely strong enough to pick up bottom material, although they may transport wave-suspended particles. Tidal range and erosive power are positively correlated. Along open areas of the Atlantic coast, the tidal range is only 1–2 m, and current strengths are normally less than 1 m/s. In constricted bodies of water such as bays and estuaries, however, the tidal range can be an order of magnitude greater, with a correspondingly greater sand-eroding ability. The world's greatest tidal range occurs in the Bay of Fundy between New Brunswick and Nova Scotia, Canada, where it reaches an astonishing 16.3 m. Current strengths of more than 5 m/s have been measured there.

Tsunami or seismic sea waves are generated by earthquakes beneath the ocean floor, by submarine landslides, or by volcanic action. Seafloor shaking generates water waves with very long wavelengths (sometimes over 500 km), wave heights of a few meters, and periods of 10 to 60 minutes. The velocity of a tsunami wave is given by

$$V = \sqrt{gd}$$

V = wave velocity (m/s)
g = gravitational acceleration (9.8 m/s^2)
d = water depth (m)

For the average ocean depth of about 4,000 m, the equation yields mean tsunami velocities of about 200 m/s (450 mph). The largest tsunami have open-ocean heights of 3–5 m and wavelengths of up to 1,000 km. Wave refraction (bending) in shallow water can transform a 5-m tsunami wave moving at 170 m/s in deep water to a wave 30 m (100 ft) high moving at 15 m/s (33 mph) at a coastline. The erosive effect of such a monstrous wave can be devastating to the coastline and to inhabitants for a considerable distance inland. The tsunami that destroyed Lisbon, Portugal, in 1755 killed 60,000 people; the Krakatau tsunami in 1883 killed 36,000.

Landward movement of the shoreline is also caused by sea-level rise, now occurring at a rate of about 30 cm/100 yr along the U.S. Atlantic coast. This rise results mostly from the melting of Antarctic glacial ice. Even greater encroachment by the sea occurs in parts of the tectonically sinking Mississippi delta, where the rate of rise can exceed 1.2 m/100 yr (Figure 14.5). Neither the melting of glaciers nor the sinking of the delta sediment is caused by human activities; however, the hypothesized increase in the greenhouse effect caused by our dumping of carbon dioxide into the atmosphere might accelerate the rate of sea-level rise by increasing the melting rate of Antarctic ice. Human activity clearly has accelerated shoreline retreat in areas such as Galveston,

Figure 14.2

Probability that a hurricane will strike a particular 80-km segment of the south Atlantic coast in a given year. *Source: R. H. Simpson and M. B. Lawrence, 1971, Atlantic hurricane frequencies along the U.S. coastline: NOAA Technical Memorandum No. NWS-SR-58, Washington, D.C.*

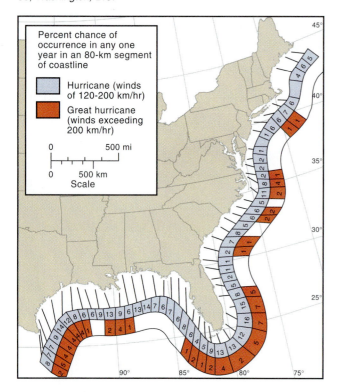

Figure 14.3

Terminology and behavior of ocean waves in deep water and as the waves approach the shore (beach). (a) Surface winds cause water molecules to move in circular orbits whose diameters decrease with increasing depth. Wave motion is negligible at a depth of 1/2 the wavelength. (b) As the waves move into shallow water—water shallower than 1/2 the wavelength—the orbits of the water molecules become compressed. The moving water scrapes the seafloor, causing erosion and sand movement. Waves increase to a height at which they cannot sustain themselves and collapse to form surf composed of very turbulent water with great erosive power.

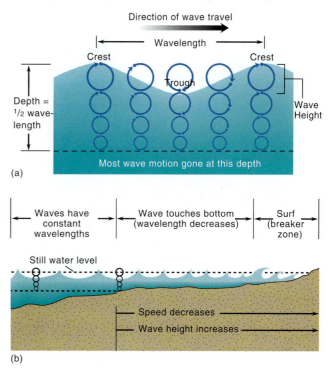

Texas, and Long Beach, California, where removal of petroleum and water from the subsurface has caused the land to subside and the sea to overrun low-lying land. In some areas river-damming has accelerated inland migration of beaches by trapping sand that would otherwise have been carried to the coast to replenish the beach.

COASTAL ENGINEERING

In a continuing effort, engineers have developed two approaches to diminishing or halting the inland movement of a shoreline. One approach includes the construction of *seawalls, jetties, groins,* and *breakwaters*. Usually, these structures are built of cement and concrete, anchored to the land and/or shallow sea floor, and designed to decrease or deflect the wave energy as it approaches the beach. Unfortunately, these structures typically create as many problems as they solve (Figure 14.6). Sand moving along the shoreface stops and accumulates on the upcurrent side of the groin or jetty, starving the beach on the downcurrent side. The beach downcurrent erodes as a result. Seawalls built at the inner edge of a beach to protect expensive buildings such as beachfront homes and resort hotels cause erosion of the very beach sand that makes the area desirable; eventually, the seawalls are undermined and disintegrate.

The second approach engineers use to maintain beaches is sand replenishment—adding "new" sand to rebuild beaches that have retreated to positions near seawalls or buildings. Sand is usually pumped to the beach from inlets, tidal delta shoals, or the continental shelf; in some cases, it is trucked from inland quarries. Beach replenishment has become more common in recent years because it does not disrupt natural processes, is a buffer against coastal erosion, and supplies sand to adjacent beaches. Unfortunately, it is also very costly and seldom lasts very long. The shortest-lived replenishment project of recent years occurred at Ocean City, New Jersey, where in 1982 storms destroyed a $5.2-million beach in only 2.5 months. As Pilkey (1989) has pointed out, "predictions of beach durability are always wrong, and nobody in the engineering community looks back to evaluate the success or failure of past projects, so no progress has been made in understanding beach replenishment."

COASTAL EROSION AND PUBLIC POLICY

During the past 25 years, state governments have become increasingly involved with property owners in such areas of public interest as coastal wetlands and shorelines. Many states have passed laws restricting development of such properties, and often these laws are so restrictive that the owners suffer great financial loss. Should the owners be financially

Figure 14.4

Wave refraction changes the wave direction, bending the wave so it becomes more parallel to shore. The angled approach of waves to shore sets up a longshore current parallel to the shoreline. *From Charles C. Plummer and David McGeary, Physical Geology, 5th ed. Copyright © 1991 Wm. C. Brown Communications, Inc., Dubuque, Iowa. All Rights Reserved. Reprinted by permission.*

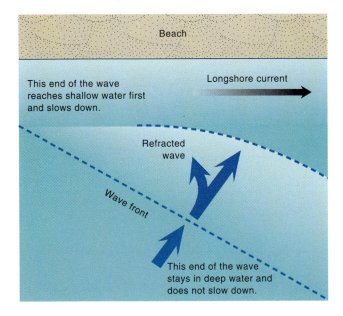

Figure 14.5

Effect of 125 years of sea-level rise in the Mississippi delta. Rapid erosion of the protecting barrier islands has exposed Louisiana's valuable wetlands and estuaries to increased storm waves and currents. *Source: S. J. Williams, et al., 1990, p. 23, U.S. Geological Survey.*

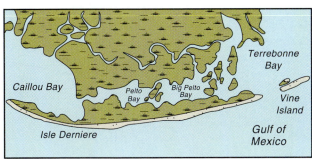

1853

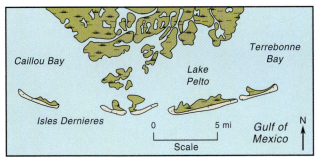

1978

Figure 14.6

Major types of engineering structures for reducing beach erosion and retreat. In all cases, accumulation of sand in one locality is balanced by sand removal (erosion) in an adjacent locality. Moving water has both an erosive capability and a sediment-transporting capacity that cannot be canceled by artificial structures. *From William H. Dennen and Bruce R. Moore, Geology and Engineering. Copyright © 1986 Wm. C. Brown Communications, Inc., Dubuque, Iowa. All Rights Reserved. Reprinted by permission of the authors.*

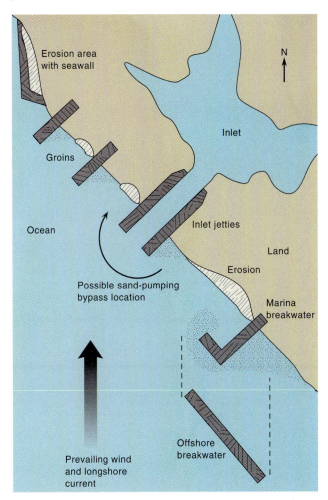

compensated? Some states claim that because property rights are not absolute rights, the state can use its police power to restrict property use to protect public health, safety, and welfare. Under this doctrine, the state is not required to pay compensation to the property owner. Property owners, on the other hand, say that the Fifth Amendment to the United States Constitution forbids the government from confiscating private property, and that declaring property unfit for commercial exploitation is equivalent to confiscation. Under this doctrine, often called the doctrine of "takings," the owners must be compensated for any financial loss the new laws cause.

The government says it cannot afford to pay for the property, but that its police power to regulate use of the property is needed to protect the environment and public safety. Property owners argue that they cannot bear the entire financial burden

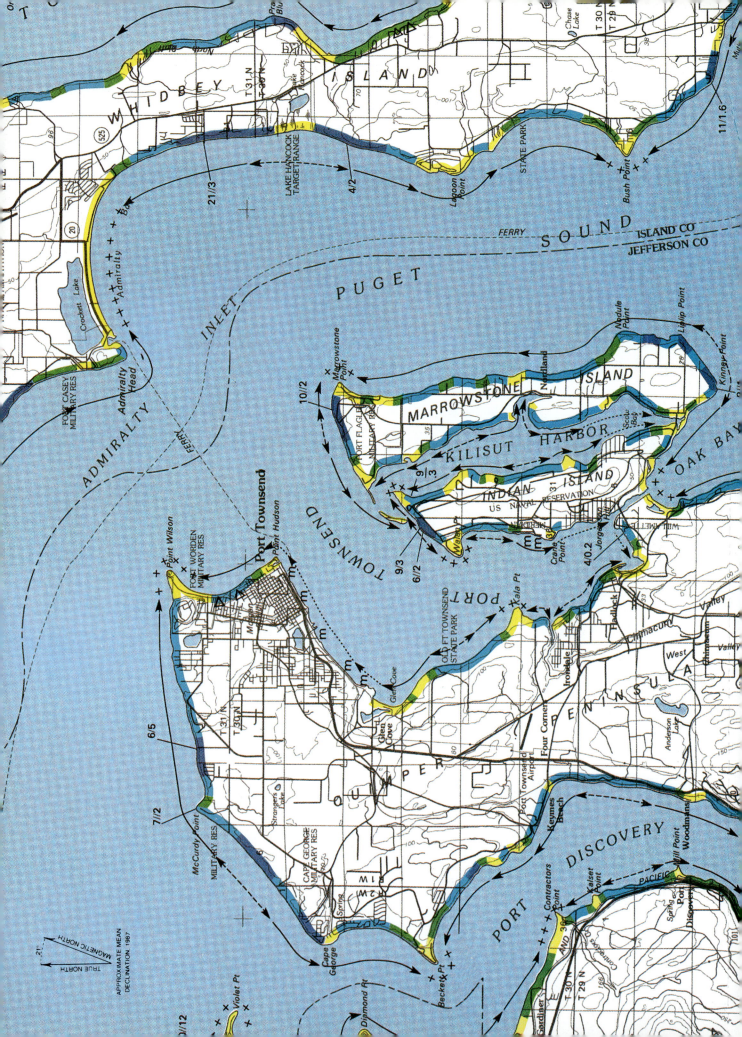

Figure 14.7

Port Townsend, Washington Quadrangle at a scale of 1:100,000 (map on opposite page).

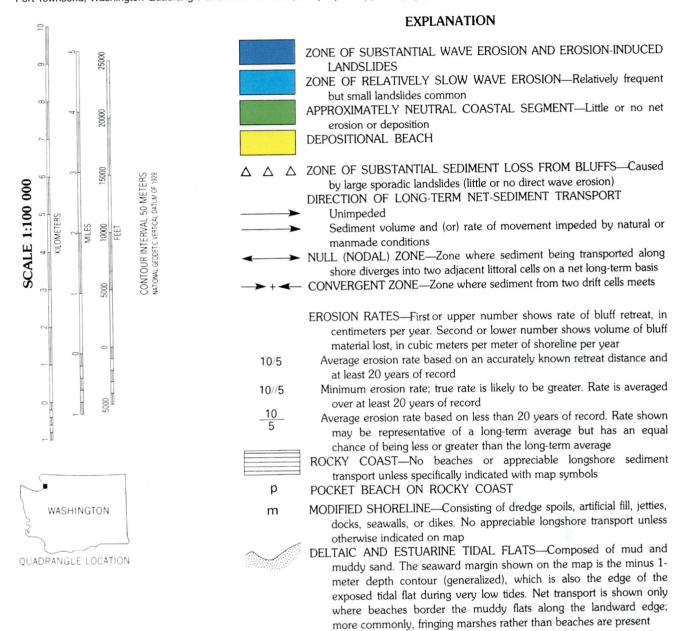

EXPLANATION

ZONE OF SUBSTANTIAL WAVE EROSION AND EROSION-INDUCED LANDSLIDES

ZONE OF RELATIVELY SLOW WAVE EROSION—Relatively frequent but small landslides common

APPROXIMATELY NEUTRAL COASTAL SEGMENT—Little or no net erosion or deposition

DEPOSITIONAL BEACH

△ △ △ ZONE OF SUBSTANTIAL SEDIMENT LOSS FROM BLUFFS—Caused by large sporadic landslides (little or no direct wave erosion)

DIRECTION OF LONG-TERM NET-SEDIMENT TRANSPORT

Unimpeded

Sediment volume and (or) rate of movement impeded by natural or manmade conditions

NULL (NODAL) ZONE—Zone where sediment being transported along shore diverges into two adjacent littoral cells on a net long-term basis

CONVERGENT ZONE—Zone where sediment from two drift cells meets

EROSION RATES—First or upper number shows rate of bluff retreat, in centimeters per year. Second or lower number shows volume of bluff material lost, in cubic meters per meter of shoreline per year

10/5   Average erosion rate based on an accurately known retreat distance and at least 20 years of record

10//5   Minimum erosion rate; true rate is likely to be greater. Rate is averaged over at least 20 years of record

$\frac{10}{5}$ Average erosion rate based on less than 20 years of record. Rate shown may be representative of a long-term average but has an equal chance of being less or greater than the long-term average

ROCKY COAST—No beaches or appreciable longshore sediment transport unless specifically indicated with map symbols

p   POCKET BEACH ON ROCKY COAST

m   MODIFIED SHORELINE—Consisting of dredge spoils, artificial fill, jetties, docks, seawalls, or dikes. No appreciable longshore transport unless otherwise indicated on map

DELTAIC AND ESTUARINE TIDAL FLATS—Composed of mud and muddy sand. The seaward margin shown on the map is the minus 1-meter depth contour (generalized), which is also the edge of the exposed tidal flat during very low tides. Net transport is shown only where beaches border the muddy flats along the landward edge; more commonly, fringing marshes rather than beaches are present

of taking the property out of circulation. The courts are currently considering this issue.

Problems

1. Draw a sketch of a coastline, with headlands (protrusions of land toward the sea), and show the change in shape of the wave front as it passes from deep water into shallow water near the headlands. Explain the reason for the change.

2. From a physics viewpoint, why do you think wave height is so important in determining wave energy? How does this relate to surfing along the California coastline?

3. Figure 14.7 is a map of the Port Townsend, Washington, Quadrangle at a scale of 1:100,000, showing the coastline about 30 miles north of Seattle. Read the legend carefully so you understand what the various symbols and colors indicate.

  a. Explain why the ratio between the linear-retreat rate and the volume of sediment eroded is so variable along the coastlines on the map. Should a site where the land is retreating more rapidly always yield a greater volume of sediment?

  b. Why is there an extensive beach at Admiralty Bay?

  c. Why is there a zone of intense wave erosion around the Lake Hancock Target Range but not at Admiralty Bay?

d. What has caused the extension of land at Point Wilson? Do you think this extension will eventually seal off Admiralty Inlet? Explain.

e. Explain why wave erosion tends to be heavy for many miles on either side of McCurdy Point. How does this tendency relate to the existence of Beckett Point?

f. Why is erosion much more intense along the coast just south of Irondale than immediately across the bay at Jorgenson Hill?

g. A large amount of sediment transport seems to be occurring in this map area. What can you infer about the character of the coastal sediments that form the source materials?

h. Based on the varying directions of sediment transport shown, the current patterns appear complex. What factors might explain this complexity?

i. Based on the topographic variations and the patterns of erosion and deposition, select the safest places to build a house. Now choose the least-safe places.

j. Global warming is causing sea level to rise. How would a rise in sea level of 50 ft affect the area where you chose to build your house? Suppose sea level rose 100 ft?

4. When examining a topographic and bathymetric map of a sandy coastline, how many ways can you think of to determine the direction of longshore current movement? In other words, what features on the ground or on the shallow sea floor might indicate current patterns?

5. Figure 14.8 shows the coastline around Morro Bay, California.

a. Which way is the longshore current flowing? How do you know?

b. What is the future of the bay southwest of the town? How do you know?

c. Suppose sea level rose 50 ft over the next 100 years because global warming caused the Antarctic ice cap to melt. Describe the consequences to the shoreline and to the communities around Morro Bay. Use a colored pencil to show the area that would be drowned.

6. You are considering buying some beachfront property but are concerned about erosion. What coastline features (rocks, sediments, and ocean) would be important to weigh in evaluating the durability of your property?

7. You own a stretch of seafront property, and some of your neighbors upcurrent are considering constructing a groin to widen the beach in front of their houses. How might this affect your property?

Further Reading/References

Dolan, R., Anders, F., and Kimball, S., 1988. *Coastal Erosion and Accretion—National Atlas of the United States.* Reston, Virginia, U.S. Geological Survey, 1 sheet.

Fletcher, C. H., III, 1992. "Sea-level trends and physical consequences: Applications to the U.S. shore." *Earth-Science Reviews,* v. 33, pp. 73–109.

Living with the Shore series. Durham, North Carolina, Duke University Press. (A continuing series of books for lay people, each book dealing with a different location along the U.S. coastline.)

Pilkey, O. H., 1989. "The engineering of sand." *Journal of Geological Education,* v. 37, pp. 308–11.

Platt, R. H., Bently, T., and Miller, H. C., 1992. "The failings of U.S. coastal policy." *Environment,* v. 35, July, p. 7–10.

Thieler, E. R., and Bush, D. M., 1991. "Hurricanes Gilbert and Hugo send powerful messages for coastal development." *Journal of Geological Education,* v. 39, pp. 291–98.

Williams, S. J., Dodd, K., and Gohn, K. K., 1990. *Coasts in Crisis.* U.S. Geological Survey Circular 1075, 32 pp.

Figure 14.8

Coastal features of part of the Cayucos, California, quadrangle. The map shows Morro Bay partly blocked by a sand bar, the delta of Chorro Creek, the community of Morro Bay, and surrounding topography. *From Norris Jones Physical Geology lab manual p. 175.*

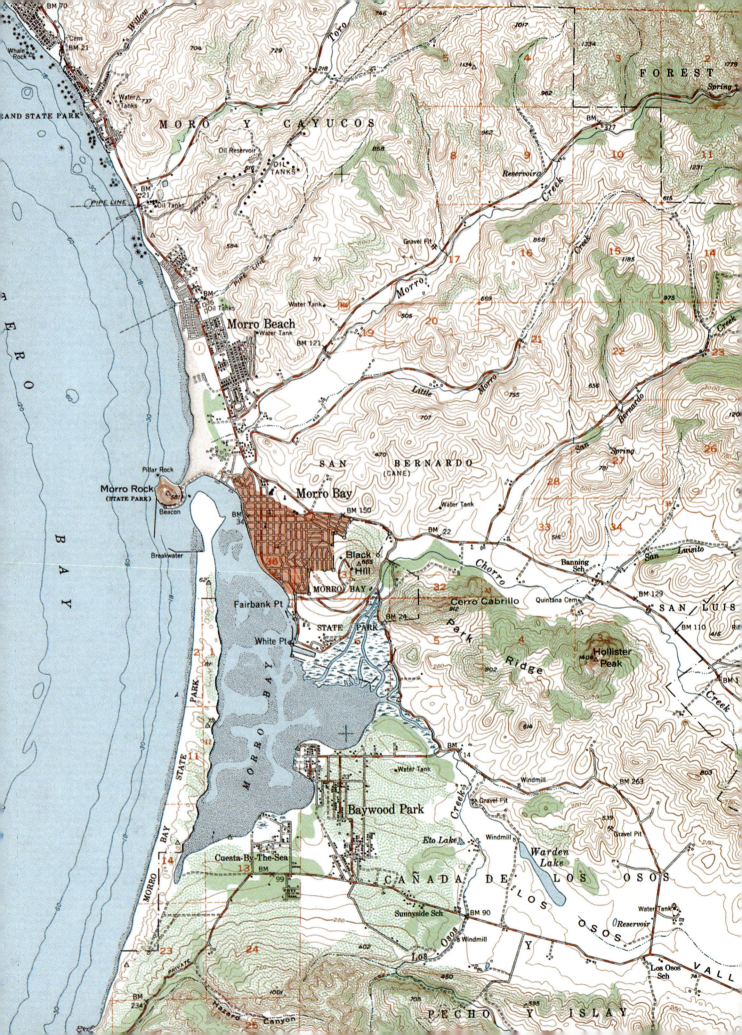

15

SEISMIC RISK AND EARTHQUAKES

For most of us, the word *earthquake* triggers visions of violently trembling ground, landslides, collapsing buildings, fires, and great loss of life. These images come from the popular press and commercial films, which give the impression that earthquakes are infrequent, but catastrophic when they do occur. In actuality, about 350,000 earthquakes occur each year, and fewer than 1,000 (<0.3%) cause noticeable damage or loss of life. Furthermore, 72% of the damaging quakes are concentrated in a narrow band around the periphery of the Pacific Ocean. Most of the rest occur in the mountain belt that extends from the Alps in southern France eastward through the Middle East to the Himalayan range in northern India and southern China. In the United States the area of most concern is the California coast, which has a high population density, two of the nation's largest cities (Los Angeles and San Francisco), and considerable topographic relief on unstable slopes. These three factors are a prescription for disaster in an earthquake-prone region.

Most earthquakes result from shock waves created by the sudden release of slowly accumulated stress in rigid bedrock. This release of stress causes blocks of rock to move along fractures called faults. The rock then becomes offset on either side of the fault. The sliding motion relieves the stress for a while, but as stress continues to accumulate, another sudden movement parallel to the fault surface is in-

evitable. This situation exists all along the San Andreas fault (actually a fault zone with many large and small subparallel faults), which runs just east of Los Angeles and slightly west of San Francisco. The slice of the earth's crust west of the San Andreas fault has been moving northward for about 40–50 million years at an average rate of roughly 0.4 in./yr. The human problem is that this movement does not occur by continuous sliding at that imperceptible rate; instead, strain energy is stored continuously, and then relieved sporadically—and unpredictably—by rapid movements of an inch or more.

MAGNITUDE AND INTENSITY

The *magnitude* of an earthquake refers to the amount of energy released, based on a scale devised by Charles F. Richter. This energy release is measured by the amount of ground displacement, or shaking, the earthquake produces. The Richter scale is logarithmic, which means that an earthquake of magnitude 5 causes 10 times as much ground movement as one of magnitude 4—and 100 times as much movement as one of magnitude 3. At increased magnitudes the amount of energy released rises even faster, by a factor of 28 for each unit of magnitude (Table 15.1a). The largest recorded earthquakes had magnitudes of about 8.9.

An alternate way to describe the size of an earthquake is by its *intensity,* a measure of the effect of the quake on

TABLE 15.1a

Frequency of Earthquakes of Various Magnitudes on the Richter Scale and the Amount of Energy Released

Description	Magnitude	Number per Year	Approximate Energy Released (ergs)
Great earthquake	over 8	1 to 2	over 5.8×10^{23}
Major earthquake	7–7.9	18	2–42×10^{22}
Destructive earthquake	6–6.9	120	8–150×10^{20}
Damaging earthquake	5–5.9	800	3–55×10^{19}
Minor earthquake	4–4.9	6,200	1–20×10^{18}
Smallest usually felt	3–3.9	49,000	4–72×10^{16}
Detected but not felt	2–2.9	300,000	1–26×10^{15}

Source: Data from B. Gutenberg in Earth, *2d ed by Frank Press and Ray Siever, 1978, W. H. Freeman and Company.*

TABLE 15.1b

Modified Mercalli Intensity Scale (Abridged)

Intensity	Description
I	Not felt.
II	Felt by persons at rest on upper floors.
III	Felt indoors—hanging objects swing. Vibration like passing of light trucks.
IV	Vibration like passing of heavy trucks. Standing automobiles rock. Windows, dishes, and doors rattle; wooden walls or frames may creak.
V	Felt outdoors. Sleepers wakened. Liquids disturbed, some spilled; small objects may be moved or upset; doors swing; shutters and pictures move.
VI	Felt by all; many frightened. People walk unsteadily; windows and dishes broken; objects knocked off shelves, pictures off walls. Furniture moved or overturned; weak plaster cracked. Small bells ring. Trees and bushes shaken.
VII	Difficult to stand. Furniture broken. Damage to weak materials, such as adobe; some cracking of ordinary masonry. Fall of plaster, loose bricks, and tile. Waves on ponds; water muddy; small slides along sand or gravel banks. Large bells ring.
VIII	Steering of automobiles affected. Damage to and partial collapse of ordinary masonry. Fall of chimneys, towers. Frame houses moved on foundations if not bolted down. Changes in flow of springs and wells.
IX	General panic. Frame structures shifted off foundations if not bolted down; frames cracked. Serious damage even to partially reinforced masonry. Underground pipes broken; reservoirs damaged. Conspicuous cracks in ground.
X	Most masonry and frame structures destroyed with their foundations. Serious damage to dams and dikes; large landslides. Rails bent slightly.
XI	Rails bent greatly. Underground pipelines out of service.
XII	Damage nearly total. Large rock masses shifted; objects thrown into the air.

people and structures. Intensity varies considerably because of factors such as local geologic conditions, quality of construction, and distance from the *epicenter* of the earthquake. The epicenter is the place on the ground surface directly above the spot where the rock ruptures. The magnitude of any particular earthquake is a constant; its intensity is a variable. Many intensity scales are in use, but in the United States the most widely used is the *Modified Mercalli Scale* (Table 15.1b).

The amount of property damage from an earthquake of a particular magnitude at a given distance from the epicenter depends mostly on the character of the underlying sediment or bedrock, as was clearly illustrated in the Loma Prieta earthquake of October 17, 1989 (Figure 15.1). Shortly before the start of a World Series baseball game in Candlestick Park in San Francisco, fans were shaken by a magnitude 7.1 quake that struck Loma Prieta 60 miles to the southeast. The main quake lasted 15 seconds but caused

Figure 15.1

Damage to structures caused by the October 17, 1989, Loma Prieta earthquake. Damage in the Marina district of San Francisco. Note the apparent lack of damage to the buildings on the next corner. Why? © *David J. Cross/Peter Arnold, Inc.*

$6 billion in damage in the San Francisco Bay area. Aftershocks with magnitudes of 5 or less continued for weeks. Most of the damage was not in Loma Prieta but in the Marina District at the tip of the San Francisco Peninsula, adjacent to the bay (Figure 15.2).

Why did most of the damage occur so far from the epicenter of the earthquake? The explanation lies in the character of the rocks and sediments of Loma Prieta and in the Marina District. At Loma Prieta the soil is underlain by hard rock. But in the Marina District the thin soil is underlain by soft mud covered by artificial fill, a mixture of 70% sand and 30% mud. Vibrations from the earthquake at Loma Prieta caused liquefaction of the muddy sediment 60 miles away near the bay in San Francisco. The ground collapsed, breaking foundations, streets, and underground utilities. The city's water system stopped functioning, so that fires fueled by ruptured gas lines spread unchecked through the district. The reason for the extensive damage was the same as it was during the famous 1906 earthquake that hit the Bay area. No doubt the same thing will happen again someday (tomorrow?) because 80% of the San Francisco penin-

sula is underlain by unconsolidated sediment. The difference in damage between Loma Prieta and the Marina District demonstrates the need to consider geologic conditions when establishing building codes and zoning ordinances in earthquake-prone areas. Zoning also should consider whether the terrain is steep or flat and evaluate the structural characteristics of existing structures such as buildings, dams, and bridges.

During the 6.7 magnitude earthquake at Northridge, California, in 1994, damage totalled $30 billion, despite the fact that Northridge rests on rock rather than unstable sediment. Northridge is a suburb of Los Angeles, and there was much more expensive construction than in the Marina District of San Francisco. A large part of the cost of repairing the damage at both San Francisco and Northridge was paid for by federal disaster loans, much of the money from taxpayers who don't live in California.

The disastrous 6.9 magnitude earthquake at Kobe, Japan, caused $100 billion in damage and incinerated the equivalent of 70 city blocks. As in San Francisco's Marina District, the cause was liquefaction of bay mud in a port

Figure 15.2

Generalized geologic map of the upper part of the San Francisco Peninsula, California. *Source: Borcherdt, 1975, p. 4, U.S. Geological Survey.*

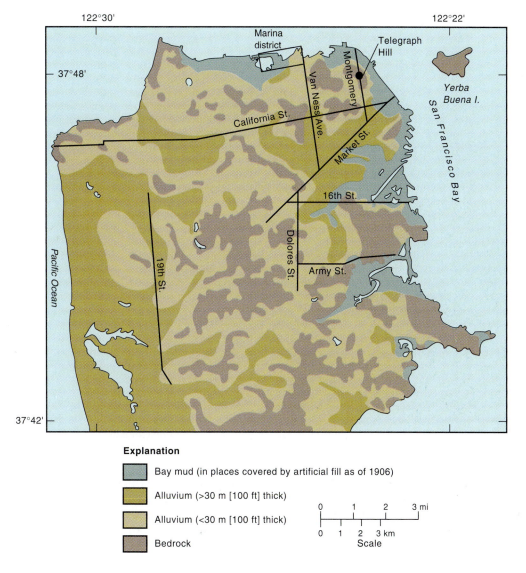

Explanation

Bay mud (in places covered by artificial fill as of 1906)

Alluvium (>30 m [100 ft] thick)

Alluvium (<30 m [100 ft] thick)

Bedrock

Scale

city. The damage in Kobe would have been even worse, were it not for "earthquake-resistant" buildings required by the city's building code since 1971. Building damage was reduced 85% by the code.

Although most quakes occur along the Pacific rim, some of the largest earthquakes in the United States during historic times occurred near New Madrid, Missouri, in 1811–1812. Damage was slight, as few people lived in the area at that time, but a similar tremor today would cause a major disaster. The major population centers of Nashville, St. Louis, and Memphis are not far from the 1811–1812 epicenters. Memphis and St. Louis are built on loose sediment, as are several other large cities in the Ohio River valley.

EARTHQUAKES CAUSED BY HUMAN ACTIVITIES

Earthquakes can be caused by human activities as well as by natural causes, as exemplified by an earthquake swarm that occurred near Denver, Colorado, between 1962 and 1965 (Evans, 1966). Since 1942, chemical-warfare products had been manufactured on a large scale at the Rocky Mountain Arsenal, about 10 miles northeast of Denver. One by-product of this operation was contaminated wastewater that, until 1961, was disposed of by evaporating it from dirt reservoirs. After the wastewater was found to be contaminating the local groundwater supply and endangering crops, an injection well was drilled so the water could be disposed of below the reach of surface processes. The well was drilled

Figure 15.3

Number of earthquakes per month recorded in the Denver area and the Monthly volume of contaminated wastewater injected into the Arsenal well. *Source: Data from Evans,* Mountain Geologist, *Volume 27, page 27, 1966, Rocky Mountain Association of Geologists, Denver, CO.*

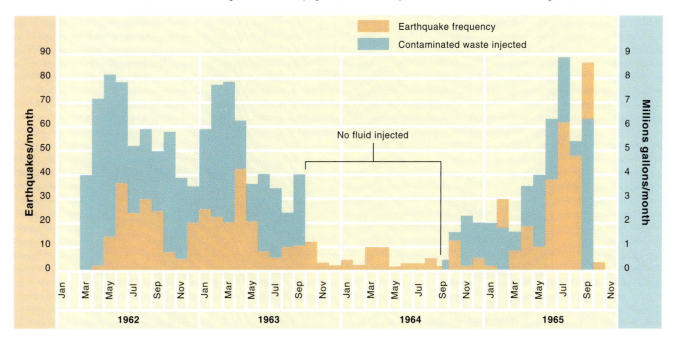

through 11,895 feet of sedimentary rocks into the gneissic basement, to a total depth of 12,045 feet, and fluid was pumped down it at rates as high as 9 million gallons per month, to a total of 150 million gallons (Figure 15.3). The result of this disposal method was that between April 1962 and September 1965, 710 earthquakes with magnitudes of 0.7 to 4.3 occurred with epicenters in the Denver area—although only one earthquake had been recorded there before, in 1892. Clearly, a strong correlation existed between the fluid injection and earthquake frequency; the injected fluid decreased the friction between fracture surfaces, triggering fault movement. Despite the federal government's repeated denials of culpability, the disposal well was eventually shut down. The earthquakes attributable to this disposal well are simply another example of a harmful interaction between humans and their natural environment.

EARTHQUAKE PREDICTION

The problem of earthquake prediction is currently receiving considerable attention in Japan, the former Soviet Union, China, and the United States, all of which have suffered significant property damage and loss of life from large earthquakes in recent years. Although a few successful predictions have been made, no reliable method of short-range prediction yet exists. The development of such a method seems many decades in the future and, even then, it is questionable whether any method will be accurate enough in view of economic incentives and human behav-

ior. For example, suppose seismologists (scientists who study earthquakes) announce that a major earthquake is "likely" to occur in May or June "in the vicinity of" Oakland, about 10 miles east of downtown San Francisco. How useful is this information? Is everyone within a 10- to 20-mile radius of Oakland likely to leave their houses for two months or more? Will offices and industries shut their doors during this period? If the "likely" quake fails to occur, will lawsuits be filed by people who have suffered financially because of the evacuation? Conversely, can seismologists be held responsible for failing to predict an earthquake that *does* occur? Such questions will become more significant in the future, as predictions of the locations and timing of earthquakes become increasingly more accurate.

EARTHQUAKE WAVES

Strong disturbances on or within the earth create vibrations that travel as waves outward from the source, similar to the effect of dropping a pebble in a lake. Earthquake waves are of four kinds: P-waves, or compressional waves, S-waves, or shear waves, L-waves, or Love waves, and Rayleigh waves. P-waves and S-waves travel through the earth. L-waves and Rayleigh waves travel along the concentric surfaces that mark discontinuities within the "shells" that form the planet.

P-waves move the fastest, S-waves are slower, and the surface waves are the slowest of all (Figure 15.4). L-waves

Figure 15.4

Travel-time curves for P-waves, S-waves, and L-waves.

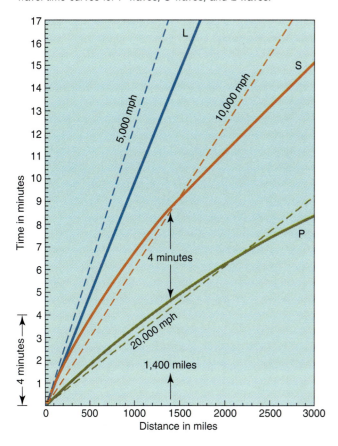

and Rayleigh waves travel at similar velocities. Analysis of the different times of arrival of earthquake waves permits seismologists to determine the epicenter of the quake.

All waves generated from an earthquake must start at the same point. However, the different types of waves travel at different speeds. The farther they have travelled, the greater the difference between their time of arrival at any

distant point and the farther the epicenter is from the recording station. The difference in time of arrival of the first P-wave and the first S-wave (Figure 15.5) is used to determine the distance to the epicenter.

A single recording station is insufficient to locate the epicenter because the distance to the epicenter can be in any compass direction from the recording station. The seismogram at the recording station can only determine the distance, not the direction. Circles from two recording stations narrow the possible epicenter to two points (Figure 15.6). It takes three recording stations to pin down the epicenter to one spot. The epicenter is the spot where the three circles intersect. Figure 15.6 shows the epicenter of the quake shown in Figure 15.5 as determined using stations in Austin, Texas (University of Texas), Ann Arbor, Michigan (University of Michigan), and Berkeley, California (University of California).

EARTHQUAKE PLANNING

Predicting earthquakes with the needed accuracy is unlikely to happen in the near future. However, we can obtain some degree of protection by building only earthquake-resistant structures in relatively safe locations. Such areas are identified by seismic risk studies. Important considerations include:

1. Does the area have a history of earthquakes during historic times? How strong? How near?
2. How thick is the soil and what is the nature of the rock or sediment that lies beneath it?
3. At what depth is the water table in relation to potentially unstable sediment?
4. Is the topography flat or hilly? Have landslides happened in the past? Where?
5. What about roads? Will there be access to the area if landslides occur and the main highway is blocked?

Figure 15.5

Arrival times at Austin, Texas for P-, S-, and L-waves from an earthquake far from Austin. S-waves normally have a larger amplitude (wave height) than P-waves. The amplitude of L-waves is much larger than either P or S.

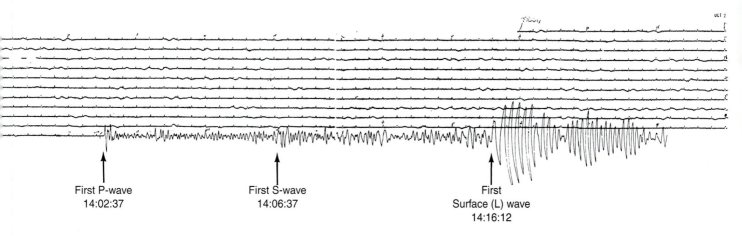

First P-wave
14:02:37

First S-wave
14:06:37

First
Surface (L) wave
14:16:12

Figure 15.6

Locating the epicenter of the earthquake shown in the seismogram from Austin, Texas. The epicenter is along the continental divide near Helena, Montana.

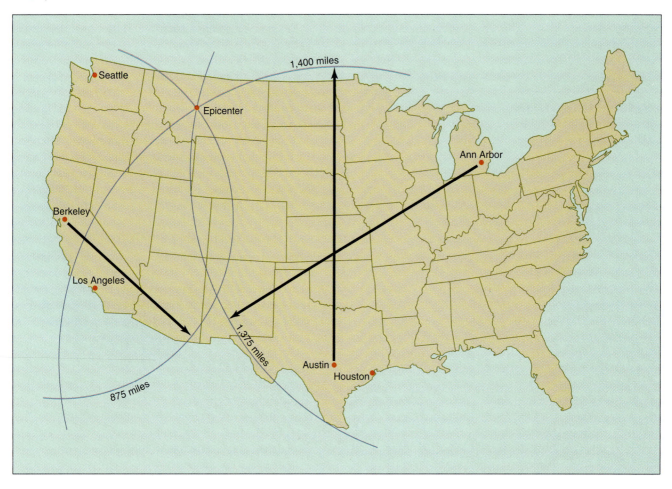

6. Where are the buried gas, water, and sewer lines? What will happen if they are ruptured?

7. How far away is help, such as the fire department or a hospital?

8. Are there toxic hazards nearby, such as chemical plants, refineries, or storage tanks for petroleum products, either above ground or buried?

9. Is there a nuclear power plant nearby and in which direction does the wind usually blow? (Yes, nuclear plants are sometimes built near historically active faults.)

A seismic risk study should be considered for every construction project in areas with a history of earthquakes during the past few hundred years.

Problems

1. The Richter scale of magnitude is open-ended, with no upper limit. However, no earthquake has been recorded with a magnitude greater than 8.9 and, as is evident in Table 15.1, only a trivial number of earthquakes have magnitudes of 8 or more. Explain these observations.

2. Figure 15.3 shows a one-year period during which no fluid was injected into the disposal well, but earthquakes occurred without interruption. How might you explain this? What might explain the less-than-perfect correlation between the amount of water injected each month and the number of earthquakes?

3. Figure 15.7 shows the numerical data for earthquake intensities during the San Fernando, California, shock (magnitude 6.4) of February 9, 1971.

   a. Construct an isointensity contour map at a unit interval for the data and give plausible reasons for the regularities and irregularities in the shapes of the contours.

   b. Construct an X-Y graph of earthquake intensity versus distance between San Fernando, where the maximum intensity recorded was 11, to Las Vegas. Interpret the shape of the curve.

4. Figure 15.8 shows the seismograms recorded at Seattle, Houston, and Los Angeles for an earthquake that occured one spring morning.

Figure 15.7

Intensity distribution map of the San Fernando, California, earthquake of February 9, 1971. *Source: Blair et al, 1971, p. 17, U.S. Geological Survey.*

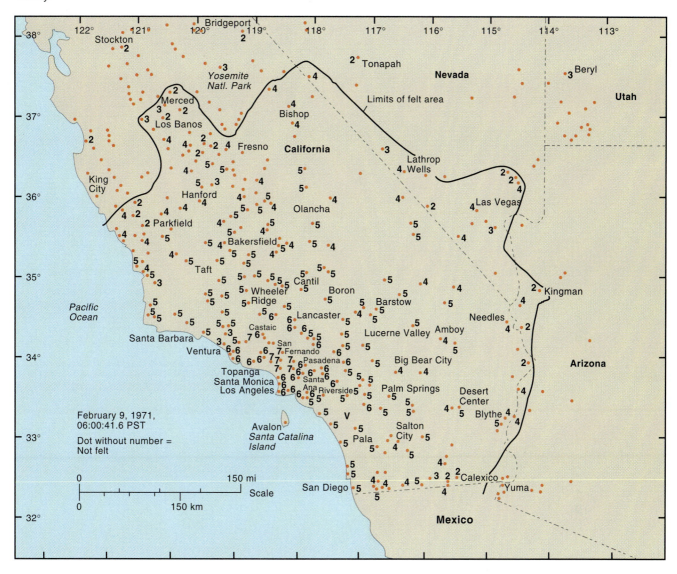

a. Which city felt the shock first? Why?

b. Which station was farthest from the epicenter? How can you tell without using a calculator?

c. All three seismograms show P-, S-, and surface waves. However, in seismograms from some earthquakes, L-waves are difficult to separate from S-waves. What is the circumstance that might cause this to occur?

d. Locate the epicenter of the earthquake recorded by the three stations and plot it on Figure 15.6.

e. You now know the distance from each recording station to the epicenter. You also know the arrival time of the P-waves at each station. Using these data and Figure 15.4, determine when the P-wave left the epicenter, that is, the time at which the earthquake occurred.

5. The San Andreas fault trends N-S up the San Francisco peninsula (perhaps along 19th Street on Figure 15.2), and the Hayward Fault runs parallel to it and 20 miles to the east, near Oakland. Seismologists have projected a 67% probability that an earthquake of magnitude 7 or more will occur along one of these faults within the next 30 years. What, if anything, should the city councils of potentially affected areas do with this information? What can insurance companies do with it? What should residents of these areas do?

6. Figure 15.9 is a generalized lithologic, isopach, and topographic map of southern Sonoma County, California, just north of San Francisco. Movement along the San Andreas fault in 1906 created offsets of as much as 12 feet in Sonoma County. Seven other potentially active faults

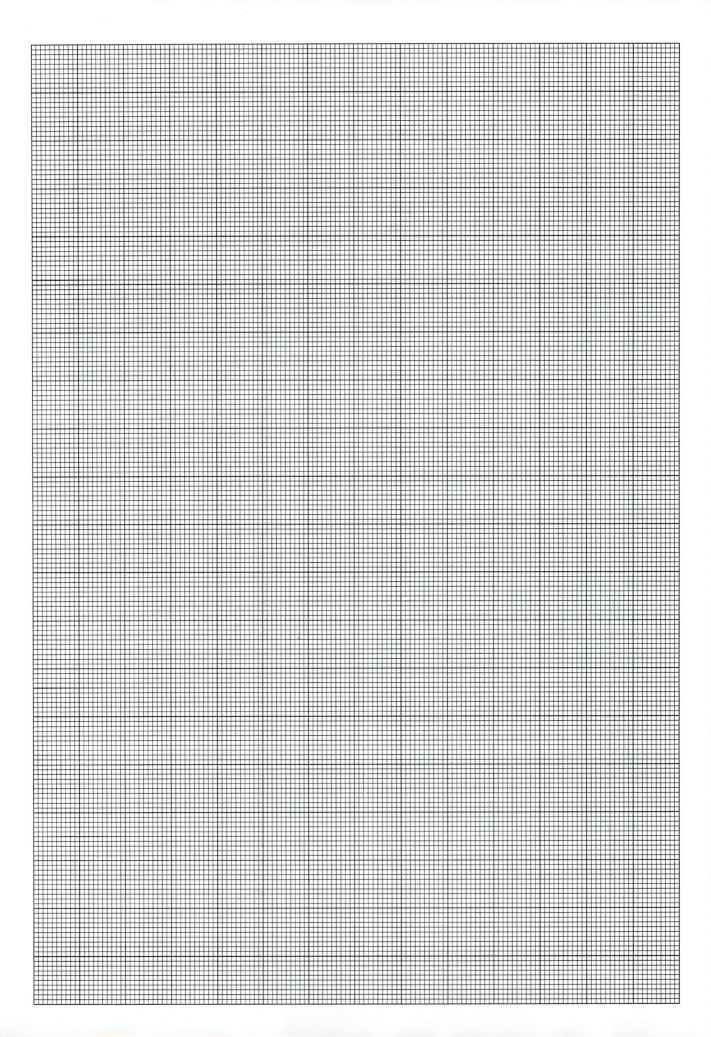

Figure 15.8

Seismographic traces of an earthquake as recorded at three locations.

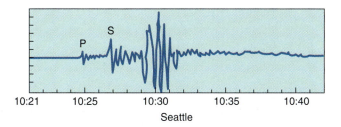

Seattle

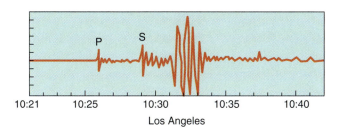

Los Angeles

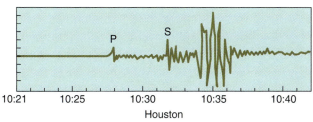

Houston

(Greenwich Mean Time)

have been identified in the area, all trending NW-SE, parallel to the bedrock outcrop on Figure 15.9. Describe how you would approach the problem of creating a seismic-risk analysis report for the residents of the area. What types of information would you like to have that are not given on the map?

Further Reading/References

California Seismic Safety Commission, 1992. *The Homeowner's Guide to Earthquake Safety.* Seismic Safety Commission # 92–02, 28 pp.

Reiter, L., 1991. *Earthquake Hazard Analysis.* New York, Columbia University Press, 254 pp.

Steinbrugge, K. V., and Algermissen, S. T., 1990. *Earthquake Losses to Single-family Dwellings: California Experience.* U.S. Geological Survey Bulletin 1939-A, 65 pp.

Weigand, Peter W., 1994. "The January 17, 1994 Northridge (California) earthquake: a personal experience." *Journal of Geological Education,* v. 42, p. 501–6.

Figure 15.9

Rock and sediment distributions and thicknesses in southern Sonoma County, California. *From R. W. Greensfelder, 1980, California Division of Mines and Geology* Special Report 120, *plate 1B. Used with permission*

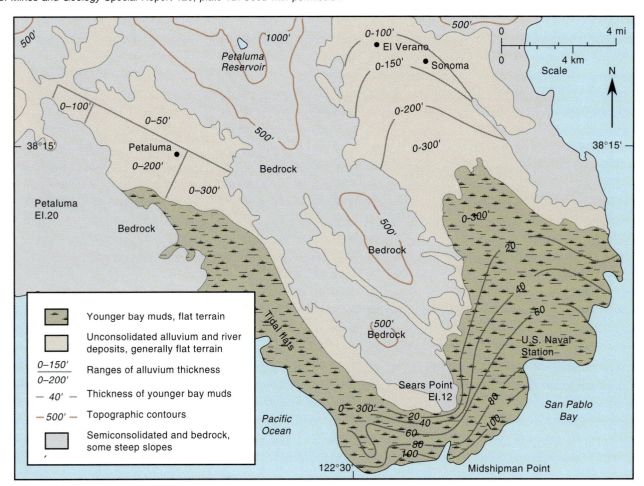

16

VOLCANOES AND ERUPTIONS

Volcanic eruptions occur infrequently and cause low annual damage compared to other hazards, such as earthquakes, floods, and ground failures (subsidence, sinkholes, landslides). In the United States, for example, the annual economic loss from volcanic eruptions is at least an order of magnitude less than the loss (0.6×10^9 1980 dollars) caused by earthquakes, which in turn is nearly an order of magnitude less than the loss associated with floods and ground failures. Nevertheless, for those who live in volcanic regions, the severity of the hazards caused by potential eruptions makes them an important consideration in land-use decisions.

Because of this fact, scientists have attempted to compile lists of high-risk volcanoes. The rating criteria for such lists include some or all of the following factors: (1) frequencies, sites, and nature of recorded historical eruptions (Figure 16.1); (2) information on recent prehistoric eruptions as inferred from mapping and dating studies (Figure 16.1); (3) known ground deformation and/or seismic events ("earthquake swarms"); (4) the nature of eruptive products, possible indicators of explosive potential; and (5) various demographic determinants, such as population density, property at risk, and fatalities and/or evacuations resulting from recorded historical volcanic disasters or crises. Unfortunately, all such lists are incomplete because geological and geophysical data for many volcanoes are inadequate. Many eruptions have occurred from volcanoes not previously considered high-risk.

Despite these uncertainties, volcanoes are commonly classified as *active, dormant,* or *extinct.* An active volcano is one that has erupted within recent history. If the volcano has not erupted within historic times—perhaps 5,000 years—but is fresh-looking and not significantly eroded, it is considered dormant, with the potential to become active again. A volcano is considered extinct if it shows significant erosion of its crest and flanks and has no recent eruptive history.

TYPES OF ERUPTIONS

Eruptions can be either explosive or nonexplosive. Explosive eruptions occur if the magma moving up the neck of the volcano is very viscous, so that most of the gases present in the magma remain dissolved. As the magma nears the surface the pressure decreases; increasing amounts of gas come out of solution, and the gas bubbles increase in size. Eventually, the pressure becomes great enough to make the gases burst out of the viscous liquid, creating the explosion that hurls great masses of magma and semiconsolidated volcanic debris from the throat of the volcano. Groundwater heated by the rising magma is also ejected in the explosion; in fact, the bulk of the fluids emitted from volcanoes is H_2O liquid and vapor.

A great deal of solid material may be ejected from a volcano and accumulate around the eruption site. The height and lateral distance that fragments attain depend on ejection

Figure 16.1

Map of Mauna Loa showing the surface distribution of lava flows in five different age categories. The notation "ka" stands for thousands of years before the year 1950. Thus, 0.75 ka = 750 years; 1.5 ka = 1,500 years; and 4.0 ka = 4,000 years. *Source: J. P. Lockwood and P. W. Lipman, 1987, Holocene Eruption History of Mauna Loa Volcano, Chapter 18 in R. W. Decker et al., editors, Volcanism in Hawaii, U.S. Geological Survey* Prof. Paper 1350.

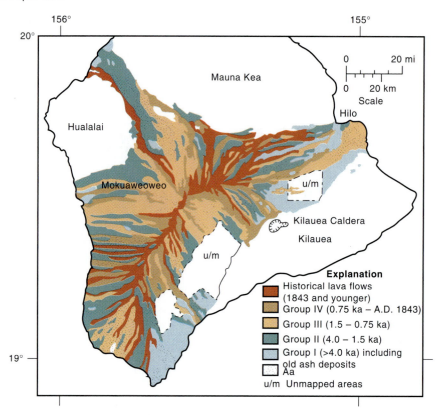

force, the size of the fragments, and wind velocity. Fragments larger than 60 mm in diameter are called bombs (round and elongate because they solidify during flight) or blocks (angular fragments); those between 60 mm and 2 mm are lapilli; those smaller than 2 mm are ash or dust. Fragments can endanger life and property at considerable distances from the volcano by forming a blanket over the ground surface and contaminating the air with abrasive particles and corrosive acids (Figure 16.2). Close to a volcano, people can be injured or killed by breathing ash-laden air; damage to property is caused by the weight of the fragments and its smothering and abrasive effects.

Glowing avalanches (*nuees ardentes,* pyroclastic flows, glowing clouds) are masses of incandescent, dry rock debris that move downslope like a fluid. They owe their mobility to hot air and other gases mixed with the debris and can travel many miles at speeds up to 100 miles per hour down valley floors on the flanks of a volcano. The path of an avalanche is guided largely by topography, but its great speed can cause it to climb vertically as much as several hundred meters, until it encounters opposing hill slopes or bends of the valley wall. This great mobility results when fragments in the moving mass are separated from each other and from the ground below by a cushion of hot, expanding gas, which largely eliminates friction as the mass moves. These flows can affect areas 15 miles or more from a volcano. Most losses from a pyroclastic flow are caused by the swiftly moving basal flow of hot rock debris, which can bury and incinerate everything in its path, and by the accompanying cloud of hot dust and gases, which can cause asphyxiation and burn lungs and skin.

Another common result of explosive eruptions is mudflows *(lahars),* masses of water-saturated rock debris that move downslope like flowing wet concrete. The debris comes from fragments of rock on the volcano flanks, and the water can come from rain, melting glacial ice and snow, a crater lake, or a reservoir adjacent to the volcano. Mudflows can be either hot or cold, depending on whether they contain hot rock debris. The speed of mudflows depends on their fluidity and the slope of the terrain; they sometimes move 50 miles or more down valley floors at speeds exceeding 20 miles per hour. Volcanic mudflows can reach even greater distances—about 60 miles from the source—than do pyroclastic flows (Figure 16.3). The chief

Figure 16.2

Map showing distribution of ash from the May 18, 1980 eruption of Mount St. Helens. Some communities were covered by as much as 3 inches of ash. *Source: U.S. Geological Survey.*

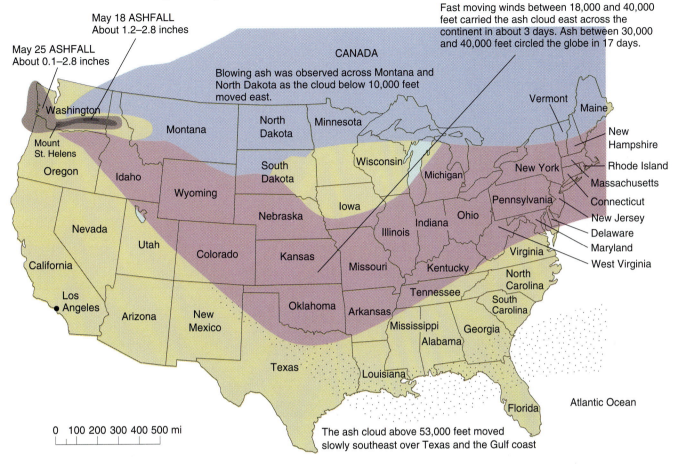

Figure 16.3

Sketch map showing the extent of two mudflows on opposite sides of Mount Rainier, Washington. The flows are confined to existing river valleys until the valleys become shallow enough for the mudflows to overflow the banks, about 25 miles from the volcano. *Source: Hays, 1981, p. 95, U.S. Geological Survey* Prof. Paper 1350.

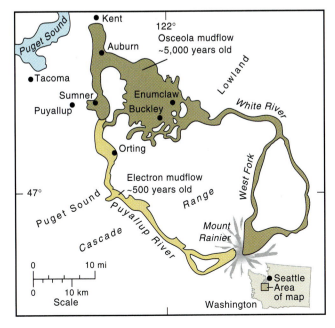

threat to humans is burial. Structures can be buried or swept away by the vast carrying power of the mudflow.

Lavas are generally erupted quietly, but can be preceded by explosive volcanic activity. The fronts of lava flows usually advance at less than human walking speed and, hence, cause no direct danger to human life (Table 16.1). Generally, however, they totally destroy the areas they cover. Lava flows that extend into areas of snow can melt it and cause floods and mudflows; lava flows that extend into vegetated areas can start fires. The diameter of the hazard zone from lava flows is normally only a few miles because the lava solidifies as it moves. Destruction is total but the affected area is relatively small.

MAGNITUDE OF VOLCANIC ERUPTIONS

The magnitude of volcanic eruptions is classified by the total amount of ejected material, both lava and pyroclastic debris (Figure 16.4). One fact evident in Table 16.2 is that nature's scales are vastly different from the human scale. The disastrous explosive activity of Mount St. Helens in 1980, for example, is of the highest magnitude on the human scale, but only a minor feature on nature's scale.

TABLE 16.1

Human Fatalities From Volcanic Activity, 1600–1986

Primary Cause of Fatalities	1600–1899		1900–1986	
Pyroclastic flows and debris avalanches	18,200	(9.8%)	36,800	(48.4%)
Mudflows (lahars) and floods	8,300	(4.5%)	28,400	(37.4%)
Tephra falls and ballistic projectiles	8,000	(4.3%)	3,000	(4.0%)
Tsunami	43,600	(23.4%)	400	(0.5%)
Disease, starvation, etc.	92,100	(49.4%)	3,200	(4.2%)
Lava flows	900	(0.5%)	100	(0.1%)
Gases and acid rain	. . .	. . .	1,900	(2.5%)
Other or unknown	15,100	(8.1%)	2,200	(2.5%)
Total	186,200	(100%)	76,000	(100%)
Fatalities per year (average)	620		880	

Values in parentheses refer to percentages relative to total fatalities for each time period. *From R. I. Tilling, "Volcanic Hazards and Their Mitigation: Progress and Problems" in Reviews of Geophysics, 27:237–69, 1989, copyright by the American Geophysical Union.*

Figure 16.4

Magnitude of volcanic eruptions.

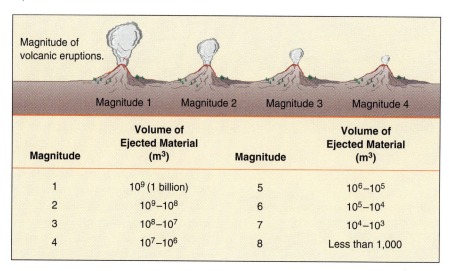

HAZARD IDENTIFICATION, ASSESSMENT, AND ZONATION

Hazard assessments exemplify a familiar geologic adage: the present is the key to the past. To make such assessments, we must assume that a volcano will probably experience the same kinds of eruptive events, in the same general areas and at about the same average frequency, in the future as it has in the past (Tables 16.2, 16.3). The danger inherent in these assumptions has surfaced frequently, sometimes with disastrous results, but we have no practical alternative. Figure 16.5 shows hazard zonation maps prepared for an area in Colombia and for the Big Island in Hawaii. Unfortunately, such maps are not yet available for many potentially dangerous volcanoes.

The incompleteness of available data poses a serious problem for public awareness of imminent danger from

TABLE 16.2

Some Famous Eruptions

Volcano	Volume New Material (billions of m³)
Crater Lake (Mt. Mazama)	42
Tambora (1915)	25
Krakatau (1883)	18
Mt. St. Helens (3,000 years ago)	8.0
Mt. St. Helens (450 years ago)	2.6
Vesuvius (A.D. 79)	2.6
Mt. St. Helens (1980)	1.6–2.0
Mt. Lassen (1914)	1.0

TABLE 16.3

Proposed Criteria for Identification of High-Risk Volcanoes

Hazard Rating	Score
1. High silica content of eruptive products (andesite/dacite/rhyolite)	
2. Major explosive activity within last 500 yr	
3. Major explosive activity within last 5,000 yr	
4. Pyroclastic flows within last 500 yr	
5. Mudflows within last 500 yr	
6. Destructive tsunami within last 500 yr	
7. Area of destruction within last 5,000 yr is >10 km^2	
8. Area of destruction within last 5,000 yr is >100 km^2	
9. Occurrence of frequent volcano-seismic swarms	
10. Occurrence of significant ground deformation within last 50 yr	

Risk Rating

1. Population at risk >100	
2. Population at risk $>1,000$	
3. Population at risk $>10,000$	
4. Population at risk $>100,000$	
5. Population at risk >1 million	
6. Historical fatalities	
7. Evacuation as a result of historic eruption(s)	
Total Score	_____

A score of 1 is assigned for each rating criterion that applies; 0 if the criterion does not apply. *Yokoyama, I., Tilling, R. I., and Scarpa, R. (1984). International Mobile Early-Warning System(s) for Volcanic Eruptions and Related Seismic Activities. Report of a UNESCO-UNEP-sponsored preparatory study in 1982–84. UNESCO, Paris.*

volcanic activity. The best that scientists can do, even with good data, is to estimate the likelihood of an eruption. Many times scientists' estimates prove incorrect, causing public skepticism regarding future pronouncements. Lawsuits might be lodged by those who either disrupted their lives for an eruption that failed to occur or lost relatives and/or property because a prediction was too imprecise. Scientists must continue their efforts to inform the public about the uncertainties inherent in predicting the future. The public tends to perceive scientists as Einstein-like figures who do not make errors. After all, if we can send someone to the moon, we should be able to do something as minor as recognizing an oncoming volcanic eruption—particularly from a volcano that has erupted before!

Land use around a suspect but currently dormant volcano poses a related problem. Should the government do more than simply inform the public about possible dangers? Should it prohibit someone from building a house on the flank of a certain volcano? Who is to decide what constitutes an acceptable level of risk? Obviously, these questions are political rather than scientific and must be decided by the residents of individual communities or states.

VOLCANOES AND CLIMATE

In addition to its immediate danger to nearby residents, an erupting volcano poses a more far-reaching and long-lasting danger. Major eruptions hurl both fine-grained volcanic dust and sulfur dioxide gas into the atmosphere (Figure 16.6), where water and oxygen rapidly convert the sulfur dioxide into sulfuric acid. In addition to forming acid rain, the sulfuric acid blocks solar radiation, resulting in lower temperatures at the earth's surface. The April, 1815, eruption of Tambora volcano in Indonesia, for example, cut sunlight by 25%, caused the coldest summer in New Haven, Connecticut, in 200 years, and caused many crop failures in the Northern Hemisphere. In central England the summer of 1816 was about 1.5°C cooler than the previous summer. This dismal weather is credited with inspiring Mary Shelley to write *Frankenstein,* and Lord Byron his poem "Darkness." A very large eruption in the near future might drastically affect crop yields and create and exacerbate food shortages in many areas, especially in the marginally productive regions where some of the world's poorest people live. Eruptions such as that of Tambora constitute a very real volcanic hazard in terms of the number of people affected. There is no

Figure 16.5a

The hazards zonation map for Nevada del Ruiz Volcano, Colombia. Although this map accurately anticipated the nature and areal extent of potential volcanic hazards and was available more than a month before the catastrophic eruption on November 13, 1985, its usefulness was negated by ineffective emergency management during the disaster. *Source: Data from Herd and the Comite de Estudios Vulcanologics, 1986, figure 4.*

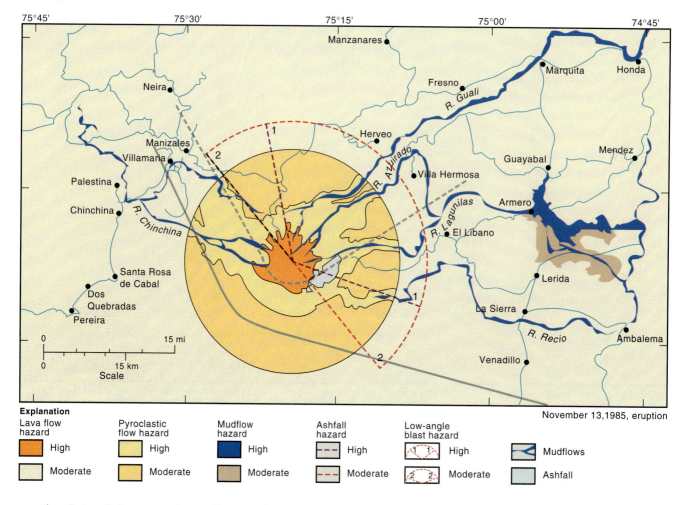

question that such large eruptions will recur; the only uncertainty lies in where and when.

Problems

1. Table 16.1 shows that the average number of fatalities per year from volcanic activity was higher in the 87-year period between 1900 and 1986 than in the previous 300 years. Explain why.

2. Examine a geologic map of the United States to answer the following questions:
   a. Where has volcanic activity occurred during the past 2 million years (Quaternary Period)?
   b. Where has volcanic activity occurred during the past 65 million years (Cenozoic Era)? Is this area larger or smaller than the area of Quaternary activity? Explain why.
   c. How does the area covered by pre-Cenozoic volcanic material compare in size with the area covered by Cenozoic activity? Explain the reason for the difference.

3. Compare Figures 16.1 and 16.5b, which show the Big Island (Hilo), Hawaii. Does the temporal sequence of lava flows in Figure 16.1 match the hazard zonation map of Figure 16.5b? What factors other than the temporal sequence of lava flows might contribute to preparation of the hazard zonation map?

4. How might you explain a difference in the size and areal distribution of fragments ejected from a volcano in different eruptions?

5. In humid climates soils form very rapidly on basaltic tephra and flow rocks after they cool. These soils are also unusually fertile. Explain these observations.

6. Basaltic lavas tend to flow quietly, whereas rhyolitic lavas tend to come to the earth's surface explosively. What might explain these tendencies? How might this difference in eruptive style be reflected in the shape of the volcanic accumulations formed from basaltic and rhyolitic lavas?

7. The 1982 eruption of the Mexican volcano El Chichón caused the emission of very large amounts of sulfur-rich gas into the atmosphere. The gas produced clouds of sulfuric acid droplets that spread around the earth. What

Figure 16.5b

Map of the Big Island showing the volcanic hazards from lava flows. Severity of the hazard increases from zone 9 to zone 1. Shaded areas show land covered by historic flows from three of Hawaii's five volcanoes (Hualalai, Mauna Loa, and Kilauea). *Source: Tilling, et al., 1987, p. 49, U.S. Geological Survey.*

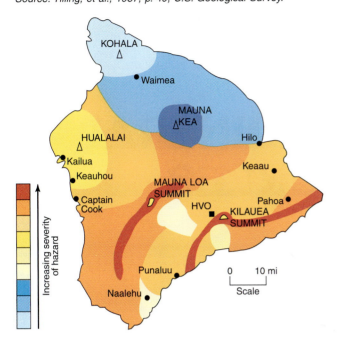

Figure 16.6a

Geologist sampling gases and measuring temperatures at Deformes fumarole (gaseous volcanic vent) on the rim of the inner crater of Galeras Volcano, Colombia. The temperature is about 500°F, and the gases contain up to 5,000 tons/day of SO_2, which is transformed to sulfuric acid in the atmosphere. The yellow material around the rim of the crater is sulfur.

Figure 16.6b

Dispersal of SO_2 gas with time from the eruption of Mt. Pinatubo in the Philippines in 1991.

Week 1

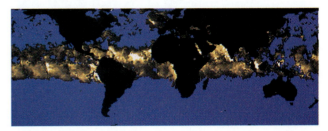

Week 6

Week 12

insurance lapse because the annual premium increased so much. You estimate it will take about two days before you are ruined financially—unless you can think of possible ways to stop or divert the lava before it enters your front door. Can you think of any schemes that might save your family home?

9. Different scientists typically have different views about the likelihood that an event of some kind will occur. Such disagreement confuses the public. How can scientists avoid confusing the public when the issue is the likelihood of a volcanic eruption?

Further Reading/References

Chester, David, 1994. *Volcanoes and Society.* New York, Chapman & Hall, 288 pp.

Peterson, D. W., 1988. "Volcanic hazards and public response." *Journal of Geophysical Research,* v. 93, no. B5, p. 4161–70.

Rampino, M. R., Self, S., and Stothers, R. B., 1988. "Volcanic winters." *Annual Review of Earth and Planetary Sciences,* v. 16, p. 73–99.

Wright, T. L., and Pierson, T. C., 1992. *Living with Volcanoes.* U.S. Geological Survey Circular 1073, 57 p.

effects do you think this sulfuric acid might have had on the earth's surface as the droplets fell to the ground?

8. You have gambled and apparently lost. You built your home adjacent to a dormant volcano, and now the volcano is starting to erupt, spilling lava from the lava lake in its crater. The lava is heading your way and you have let your

17

NONFUEL MINERAL RESOURCES

Consider your pencil. How many materials from the earth are needed to manufacture it (Figure 17.1)? The raw materials include metals such as zinc, copper, and iron, and nonmetals such as clay, graphite, and petroleum. The source of these materials is the mineral resources of the earth. As someone has said, "If it can't be grown, its gotta be mined."

A mineral *resource* is any mineral that has value to people and can be extracted from the earth *at a profit* using existing technology. The resource can be metallic (copper, gold, lead, and many others) or nonmetallic. Although metals get the most publicity, about 94% of our consumption of mined minerals is nonmetallic, mostly unglamorous materials such as crushed rock, stream gravel and sand, and limestone (Figure 17.2). These construction materials dominate mineral and rock mining in the United States. Each new home uses an average of 200 tons of crushed rock, sand, and gravel. Foundations, concrete blocks, bricks, mortar, and roofing shingles all require sand.

NONMETALLIC MINING

Nonmetallic materials are taken from the earth at a quarry. Rocks such as limestone and dolostone are blasted loose with explosives, crushed to the size wanted by the consumer (nearly always a construction company) and sold (Figure 17.3). These carbonate rocks account for 71% of the crushed rock (called aggregate) produced. Igneous rocks such as granite and basalt account for another 22%. The remaining 7% is sandstone and various metamorphic rocks.

Rock is heavy, weighing about 2.6 times the weight of the same volume of water. Transporting it to construction sites requires a fleet of large heavy-duty trucks, and the transportation cost is about 10 cents/mile/ton. Because lots of aggregate, sand, and gravel are needed in large construction projects, quarries must be located within a few tens of miles from the place where they will be used. As a result, quarries are numerous. About 6,000 different companies own 9,300 rock quarries in the United States, an average of 186 quarries per state. Almost all of them are surface operations because of the greater cost of subsurface operations.

Rock quarries are relatively benign from an environmental point of view. They tend to be small because of the need to be close to a construction site, and normally these sites are scattered widely. They produce some dust from blasting activities, increase the noise level when blasting occurs, and increase the local traffic flow. But these are small concerns when matched against the benefits the aggregate provides to a growing community. Without the stone there would be no growth.

Nevertheless, the quarry provides the seeds of its own destruction. As the community grows, it expands toward the quarry site and eventually surrounds it. And who wants a quarry as a next door neighbor? What normally happens is that the town council receives a request from concerned citizens to rezone the quarry area from industrial to commercial or residential, which is approved, and the quarry operators are given perhaps a year to shut down. The city, which

Figure 17.1

Mineral resources necessary to make a wooden pencil.

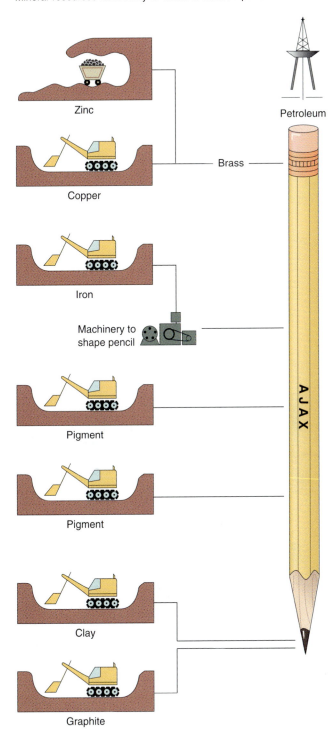

has been able to grow because of the cheap local source of construction materials, kills the industry that made its growth possible. Such matricide is common in areas around large cities such as New York and Los Angeles.

DIMENSION STONE

A few limestone deposits consist of thick, structureless beds without fractures or other partings (Figure 17.4). Such de-

posits yield blocks that can be sawed, smoothed, turned, or carved to make buildings or monuments. Blocks are carefully cut by channeling machines and loosened from the quarry floor. From there they are taken to a mill, where they are fashioned according to specifications from an architect or designer. Such dimension stone is used as construction blocks, as facing on building exteriors, and as floors, panels, and windowsills. Many other types of igneous, sedimentary, and metamorphic rocks are also quarried for use as dimension stone. As with limestone, ease of quarrying, location near a large city and available transportation, and physical attractiveness determine commercial value. The more attractive and desirable the stone, the less critical is the location. Some types of dimension stone are transported many thousands of miles because of their desirability; well-known examples are Italian marble and travertine, an unusual variety of limestone.

SAND AND GRAVEL

High-volume, high-velocity streams are the major environments of transportation and deposition of commercial gravel deposits. Fast-moving streams act as grinding and washing mills, destroying soft, structurally weak, and commercially undesirable fragments such as shale and schist and concentrating hard, sound ones such as quartz. Within a fluvial complex, commercial sand and gravel deposits are most likely either in braided streams or in the faster-moving, channel-center waters of nonbraided streams.

Most commercial sand and gravel deposits occur in one of two types of settings. In glaciated areas of the central and eastern United States, suitable deposits are abundant just south of the farthest advance of Pleistocene ice masses. Here, the sand and gravel consists of debris that was carried by the glacier, dropped as the ice melted, and then transported outward by the meltwater, the transportation serving to remove commercially undesirable silt and clay. In unglaciated areas the best deposits are located adjacent to highlands such as the Sierra Nevada in California or Rocky Mountains in Colorado (Figure 17.5). Gravels are seldom transported far from their source because of the sharp decrease in stream velocity as streams leave the mountains. Because of both its mountain ranges and its rapid growth, California produces more sand and gravel than any other state.

THE NEED FOR GEOLOGIC MAPS

Geologic maps are an indispensable tool for locating rock quarries or sand and gravel sites. Using these maps, the resource geologist locates areas where the desired rock types are exposed at the surface (or under the soil cover). Then, rock samples are collected and analyzed in a laboratory to determine the quality of the resource. For example, for limestone an important factor is its purity. Limestones that contain sand are structurally weaker, and certain chemical impurities can make the rock unsuitable for some uses. For all quarried rocks the presence of natural fractures such as parallel joints is desirable. The more

Figure 17.2

The annual per capita consumption of nonmetallic and metallic mineral resources for the United States is more than 18,000 pounds. About 94% of the materials used are nonmetallic. (*Source: U.S. Bureau of Mines*)

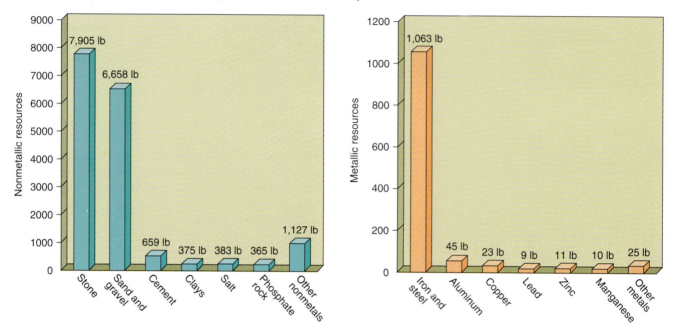

Figure 17.3

Limestone quarry in the Ozark Mountains, Arkansas.

Figure 17.4

Quarrying Indiana limestone. The straight cuts are made by channeling machines (beyond the derrick). Long blocks, freed at the bottom, are "turned down" and split into smaller blocks for removal to the finishing plant.

natural regular fractures, the easier and less costly it will be to excavate the rock.

Assuming the limestone or sand is of suitable quality and is widespread enough to last for some years, a permit to begin operations is requested from the city and the land is purchased from the current owner (Figure 17.5). A new quarry has been born.

METALLIC MINING

Minerals from which valuable metals can be extracted are called *ore minerals*. Some important minerals occur in pure form, such as gold and platinum, but in most ore minerals the valuable metal occurs in combination with other elements. Most commonly, the accompanying elements are oxygen or sulfur. Examples include galena (lead sulfide, PbS), the main ore mineral of lead for your car battery; and hematite (iron oxide, Fe_2O_3), the source of America's iron for the manufacture of steel. Some ore minerals are combinations of three or more elements, such as chalcopyrite (copper iron sulfide, $CuFeS_2$), the source of copper for home wiring. A more complete list of ore minerals is given in Table 2.5.

PLACER DEPOSITS

Metals that occur in pure form need little processing. For example, gold commonly occurs as segregations in stream sands, accumulations called *placer deposits*. The gold particles are concentrated into distinct yellow layers among the clear quartz sand grains because of their high specific gravity, 19 compared to only 2.6 for quartz. The gold drops out as stream velocity decreases, but most of the quartz keeps moving downstream. However, these segregations depend on the size of the grain (its volume) as well as its specific gravity, so that a large grain of quartz will be deposited with a small grain of gold. Hence, gold placers are not 100% gold and some processing is required to separate the gold from the quartz. Also, in commercial deposits large dredges are used to scoop out large volumes of stream sediment at a time and fairly pure gold layers become mixed with abundant quartz.

HARD-ROCK MINING

Most of the time, ore minerals occur as scattered grains in a large body of rock such as granite, and separating them from the worthless rock is an expensive process. In a surface mine, soil and other rocks that may lie above the ore-bearing rock are removed. Then the host rock, for example granite, must be attacked with explosives to produce large blocks of rock. These are taken to a mill where they are crushed and chemically treated to separate the desired mineral from the host rock. Next, the separated mineral grains are taken to a smelter where heat separates the pure metal from the oxygen or sulfur with which it is combined. In a subsurface mine the procedures differ only because vertical passageways must be created and rooms excavated for rock removal to proceed. The price of the resulting pure metal on the open market depends on its concentration in the ore-bearing rock or sediment, the cost of processing the bulk material, and supply and demand factors. Prices of commercially traded metallic elements change almost daily to reflect these variables. At present, gold and platinum sell for about $350 an ounce, silver $5, mercury $0.14, copper $0.06, and tungsten $0.03. If some of these prices seem low to the point of being trivial, keep in mind that these metals are very heavy (high atomic weights). You don't get very much of the stuff in an ounce.

ENVIRONMENTAL CONCERNS IN METAL MINING

Mining and processing of metallic ore deposits probably creates more environmental problems than any other human activity. These problems include

1. Some of the largest holes in the ground ever created (Figure 17.6) and an enormous amount of waste rock. In most metal mining, 80–90% of the excavated rock is waste. Federal law requires that the land surface be restored as closely as possible to its original condition when mining activities are completed.

2. Another serious problem resulting from mining activities at the earth's surface is the production of

Figure 17.5

Sand and gravel planning map for part of the Boulder, Colorado, urban area. *Source: USGS Circular 1110, p. 33.*

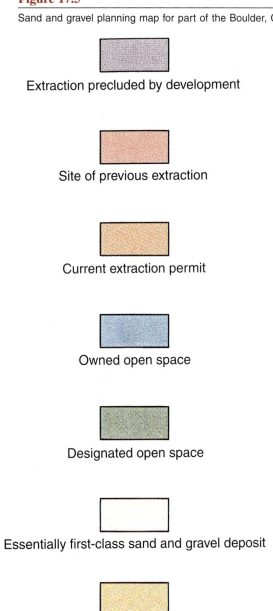

Extraction precluded by development

Site of previous extraction

Current extraction permit

Owned open space

Designated open space

Essentially first-class sand and gravel deposit

Potentially extractable lower class sand and gravel deposit

1,000 feet

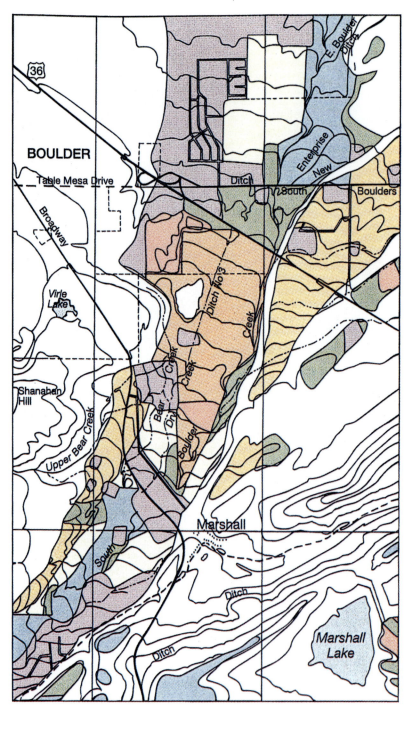

Figure 17.6

The Bingham Canyon copper mine near Salt Lake City, Utah. This open-pit mine is over 2 1/2 miles wide at the top and half a mile deep.

acid waters from the mining of metallic ores and coal (Figure 17.7). The culprit in both cases is pyrite, FeS_2. The coal mined extensively in the central Appalachians contains an average of 2.5% pyrite. The igneous rocks mined in the western United States for the economically valuable metals they contain also contain abundant sulfur. The sulfur occurs not only in pyrite. Many of the metallic ore minerals are rich in sulfur; the copper ore mineral chalcopyrite ($CuFeS_2$) and the lead ore mineral galena (PbS) are examples.

During mining, rock is broken and crushed, exposing fresh rock surfaces. Sulfur-bearing minerals in the fragmented rocks react with water and oxygen in the atmosphere to produce sulfuric acid. The acid drains into nearby streams. The acidity of stream water adjacent to mines may be 100,000 to 1,000,000 times more acidic than normal stream water. Few plants can survive such acidic waters. The acidity is also lethal for many water-dwelling animals as well—fish, snails, and others. The acid waters also sink through the soil and enter the groundwater in the mining area. For example, the Bingham copper mine in Utah, the world's largest, has released a plume of sulfuric acid water that now contaminates 50–70 square miles around the mine, giving well water a taste you wouldn't want to experience.

Figure 17.7

Acid mine drainage near Frostburg, Maryland. The orange color results from the presence in the water of perhaps 1% hematite (iron oxide). Harmful elements such as lead, arsenic, or cadmium may also be present.

3. Smelting produces solidified molten rock called *slag*. The slag pours from the smelter as the valuable metal is removed from the treated rock. Slag looks like frothy volcanic rock and normally is enriched in potentially harmful heavy metals that can leach into soil and water supplies.

4. The operation of a smelter releases sulfur dioxide gas and toxic heavy metals from smokestacks. As with the sulfur dissolved from minerals at the mine, the sulfur from smokestacks is converted in the air to sulfuric acid, acid rain that moves downwind from the smelter.

Problems

1. Suppose you are thinking of opening a quarry to serve the cemetery headstone industry. Which types of rock would be most suitable and why?

2. Do you think an iron-containing limestone would make a good building stone for a monument? Explain your reasoning.

3. One type of foliated metamorphic rock was once widely used as roofing material on private homes. Why do you think it was selected? Why do you think it is not widely used today?

4. Using the topography and geography of the sand and gravel planning map of Boulder, Colorado,
   a. Calculate the area of sand and gravel in each of the six areas labeled "first class".
   b. If it requires 200 tons of this material to build a house, how many houses can be built from these six deposits? Assume the deposit to be soft thick. (The specific gravity of the grains is 2.6 and the deposits have a porosity of 30%. Water weighs 62.4 pounds per cubic foot.)
   c. It costs $0.10/mile/ton to transport the sand and gravel by truck from a quarry to Boulder. Will the cost of transporting these materials be a significant factor in the cost of the house?
   d. Why do you think the area around Boulder is so blessed with natural construction materials?

5. Examine Table 2.5 and calculate the percentage of major ore minerals that contain sulfur as an essential element.

6. Examine the Periodic Table of the Elements in your textbook and find the atomic weight of each element. Calculate the percentage of sulfur in chalcopyrite, the major ore mineral of copper. Suppose an igneous rock contained 0.5% chalcopyrite as the only sulfur-bearing mineral. What would be the percentage of sulfur in the rock?

7. Look around your campus and identify the types of natural stone you see used for building blocks, facing stone, or decorative purposes. In the space below, list each type of rock, its location on campus, and the use to which it is being put.

Further Reading/References

Barnes, J. W., 1988. *Ores and Minerals: Introducing Economic Geology.* Philadelphia, Open University Press, 181 pp.

	Type of Rock	Location	Use
1.			
2.			
3.			
4.			
5.			
6.			
7.			
8.			
9.			
10.			

Gustin, Mae S., Taylor, George E., Jr., and Leonard, Todd L., 1994. "High levels of mercury contamination in multiple media of the Carson River drainage basin of Nevada: Implications for risk assessment." *Environmental Health Perspectives,* v. 102, p. 772–78.

Hannibal, J. T., and Park, L. E., 1992. "A guide to selected sources of information on stone used for buildings, monuments, and works of art." *Journal of Geological Education,* v. 40, p. 12–24.

Langer, William H., and Glanzman, V. M., 1993. *Natural aggregate: Building America's future.* U.S. Geological Survey Circular 1110, 39 pp.

Poulin, R., Pakalnis, R. C., and Sinding, K., 1994. "Aggregate resources: Production and environmental constraints." *Environmental Geology,* v. 23, p. 221–27.

Prentice, J. E., 1990. *Geology of Construction Materials.* New York, Chapman & Hall, 202 pp.

U.S. Bureau of Mines, 1985. *Control of acid mine drainage.* U.S. Bureau of Mines Information Circular 9027, 61 pp.

18 EXERCISE

PETROLEUM AND NATURAL GAS

Petroleum and natural gas supply about 65% of this nation's energy consumption (petroleum 40%, natural gas 25%), a decline from about 80% twenty years ago. The proportion of our petroleum that is imported has risen considerably during the same period, from about one third to nearly one half.

Petroleum and natural gas are part of a group of substances called *fossil fuels,* fuel resources that require millions of years to form naturally but are used by humans at rates high enough to exhaust the supply within the next few centuries. Our supply in the United States will be exhausted much sooner because exploration efforts have been going on for a longer time (Figure 18.1). We have already found our easily discovered petroleum resources, making further discoveries more difficult and more expensive compared to those of many other countries. As a result, we import about half of the oil we use. But in what geologic settings do petroleum and natural gas occur? How do we go about finding deposits located many thousands of feet below the surface?

Petroleum and natural gas are fluids formed largely from the organic tissues of microscopic marine organisms. These organisms live at the ocean surface and are distributed around the world by currents. After they die they settle to the ocean floor and are buried, and their organic matter is converted to liquid and gaseous *hydrocarbons* by processes not fully understood. Hydrocarbons are chemical compounds composed mostly of carbon and hydrogen. Depending on the temperatures reached during and after the conversion, liquid, gas, or both may be formed. Because the original organisms are so small, their skeletons are deposited with other fine-grained

materials such as clay-rich muds. Black muds rich in organic matter are the rocks in which petroleum and natural gas form. The fluid hydrocarbons are squeezed from the lithified muds (shales) into porous, permeable sandstones and carbonate rocks that overlie them in the sedimentary rock column. The

Figure 18.1

An oil drilling platform in a Louisiana wetland (swamp). Is the wetland being harmed?

Figure 18.2

Major types of subsurface traps for petroleum and natural gas. (a) anticline; (b) fault; (c) salt dome; (d) unconformity; (e) facies change or pinch-out. Each trap has a porous and permeable reservoir rock with an impermeable barrier above it. In (c) the barrier is a finger of salt that has risen from a salt layer below; in the other cross-sections, the barrier is a shale unit. The petroleum is overlain by a less dense gas accumulation. Below each petroleum zone is salt water, which is denser than the petroleum.

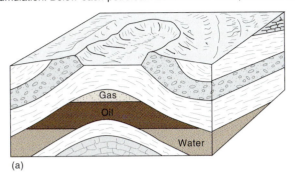

(a)

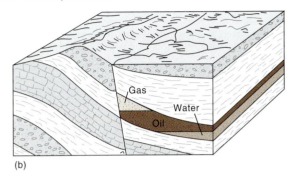

(b)

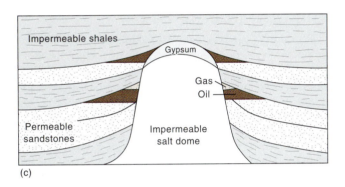

(c)

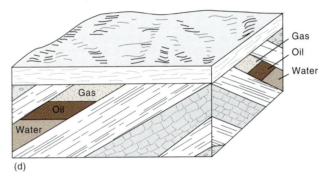

(d)

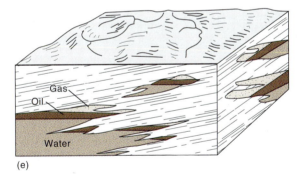

(e)

hydrocarbon liquid and gas then move through these rocks toward zones of lower pressure until stopped (trapped) by some type of change in geologic conditions.

Five types of traps are responsible for nearly all our petroleum and natural gas reserves: anticlines, faults, salt domes, unconformities, and facies changes (Figure 18.2). No matter which type of trap causes the hydrocarbon accumulation, we can recover no more than half the accumulated amount with our current technology; often we can bring only 20–30% to the surface. As a result it is correct to say, as some politicians do, that there is more oil still in the ground than has ever been removed. Unfortunately, however, we are unlikely to recover the bulk of this remaining oil because of both the high cost of developing new technology and the immense amount of low-cost oil available overseas.

RESOURCE TERMINOLOGY

Numerous terms are used in the petroleum industry and the popular press to describe amounts of hydrocarbons and evaluate available fuel resources.

1. Amounts of petroleum are normally measured in barrels (bbl), where one barrel contains 42 gallons. The standard unit of energy used in petroleum-industry literature is the British thermal unit (Btu); one gallon of petroleum generates about 138,000 Btu. Lighting a 100-watt lightbulb for one hour requires 341.5 Btu. Amounts of natural gas are measured in cubic feet. In terms of Btu energy equivalents, 6,000 ft^3 of gas equals one barrel of oil. One thousand

cubic feet of gas heats a typical home for one winter day.

2. *Reserves* are the part of a petroleum accumulation that could be economically extracted at the time the estimate is made. But the prices of petroleum and natural gas fluctuate significantly and can change precipitously as international political currents change. For example, before the Arab oil embargo in 1973, the price of a barrel of oil was about $4; by 1980 the price had reached a high of $41. In 1987 a low of $8 was reached and, since then, the price has climbed erratically up to about $20. Imagine the effect this range from $4 to $41 has on reserve estimates! Unless the estimate is based on a specific price per barrel of oil, its value can be grossly in error.

3. *Secondary recovery.* Petroleum and natural gas are like artesian water in that the fluid rises into the drill hole because of pressure release. The oil might rise all the way to the surface, forming a gusher, or might require pumping part of the way—neither of which requires pumping anything into the well. In secondary recovery, however, oil is forced from the reservoir rock by pumping water down the hole. This process is also called *enhanced recovery.*

4. *Tertiary recovery* refers to forcing oil from the reservoir pores by methods more exotic than water injection. Examples include the injection of steam, carbon dioxide gas, detergents, or bacteria. As the price of petroleum increases, the use of these exotic procedures becomes economically feasible.

ENVIRONMENTAL PROBLEMS

All exploration, production, and purification processes involving hydrocarbons have serious pollution problems. Oil is no exception. When a hole is being drilled, mud is circulated in the hole to cool the drill pipe and diamond-studded drill bit. As the mud returns to the surface it is discarded into a waste pond near the well site, from where it commonly seeps into the soil and groundwater. When the crude oil (petroleum) is refined, large amounts of sulfur and other noxious chemicals are produced. These chemicals are commonly dumped into an adjacent waterway or pumped into the ground through disposal wells. Groundwater is often contaminated by this procedure. During refining of the crude oil, sulfur and potentially harmful heavy metals are released into the air. Finally, withdrawal of petroleum from the subsurface has resulted in some areas in large-scale sinking of the ground because some of the support for overlying rocks has been removed. (Removal of groundwater can also cause subsidence.)

One well-studied example of subsidence caused by the withdrawal of petroleum occurred at Long Beach,

California. Subsidence was first noted in 1940, and by 1974 the ground surface had dropped nearly 30 feet in the central area (Figure 18.3). Although water was pumped into the ground in an attempt to halt or reverse the subsidence, only minor recovery occurred.

SPILLS FROM OCEANIC TANKERS

Petroleum from many sources has been entering the ocean for at least a billion years; it is not just a 20th-century phenomenon. The present rate of natural seepage from the sea floor totals millions of tons per year (one ton equals about 250 gal of petroleum). Even so, the oceans are not covered with a sheet or even a sheen of oil, nor are beaches heavily covered, although many beaches have small quantities of oil. Clearly, natural processes must make the oil that seeps into the oceans disappear fairly rapidly. These processes include evaporation, oxidation, bacterial degradation, and dispersal by winds, currents, and tides.

Currently, however, our civilization faces a problem: the amount of oil added to the ocean is increasing steadily because of the increasing use of large tanker ships to transport imported oil to major consumers, particularly the United States, Japan, and western Europe. Tankers now reach lengths of 1,000 ft and hold up to 2 million barrels of oil (84 million gal), so that even a single spill is likely to be a major disaster. This situation has led to considerable study of both the effects of oil pollution on the marine environment and cleanup methods for open-sea and nearshore spills. Current cleanup methods include using booms to control the lateral spread of the spill and "vacuum cleaners" and absorbents to remove the oil. Burning, sinking, and dispersing spills have also been attempted, but with mixed results. In extreme cases of beach pollution, people have been employed to clean gravel, stone by stone, along miles of beach. Such extreme cases highlight the need for developing new, automated methods for dealing with the tanker spills that are inevitable in our industrial civilization.

Problems

1. Some faults are excellent structural traps for oil and/or gas while others are not. Explain why such differences can occur. How might a "leaky" fault help oil-exploration efforts?

2. Explain how the texture and mineral composition of a sand or carbonate sediment affect its ability to serve as a future reservoir for oil and gas.

3. Figure 18.4 shows a township in the Public Land Survey System, with elevations above sea level given for points at the top of a porous, permeable sandstone unit. The Student Petroleum Exploration Club has drilled the 38 holes and discovered 11 producers, shown as black dots. The 27 dry holes are marked with the standard symbol.
   a. Using a 100-ft contour interval, construct a structure contour map of the top of the producing sandstone.

Figure 18.3

Aerial photograph of the coastal area around Long Beach, California. Withdrawal of petroleum from the Wilmington oil field resulted in ground subsidence of 30 feet.

Figure 18.4

Elevation (above sea level) of the top of a porous and permeable sandstone.

b. What type of structural trap is present?

c. How thick do you estimate the oil-producing zone to be?

d. If the sandstone is 500 ft thick and has a porosity of 15%, how many barrels of oil might the reservoir contain?

e. There is some acreage for lease in the center of section 15. Do you think it is a good bet to lease this property and drill? Explain your decision. How about property in the center of section 16?

4. Most ground subsidence occurs in areas underlain by fragmental sediments deposited within the past few tens of millions of years (the last half of the Tertiary Period). Explain why this is so.

5. Shown below are data from the Inglewood oil field in Los Angeles County, California. The oil-producing horizon is of Pliocene and possibly Pleistocene age and consists of poorly consolidated marine silts and very fine-grained sands 1,000–2,000 ft deep. The table indicates the amount of liquid produced (oil plus associated salt water) and the

volume of land subsidence for five time periods since oil production began in 1911.

	Liquid Production (ft³)	Volume of Subsidence (ft³)
Nov. 1911–Oct. 1943	1,130,000,000	95,552,000
Oct. 1943–March 1950	567,600,000	39,900,000
March 1950–Aug. 1954	433,000,000	35,860,000
Aug. 1954–Oct. 1958	376,600,000	27,020,000
Oct. 1958–Aug. 1962	299,800,000	19,760,000

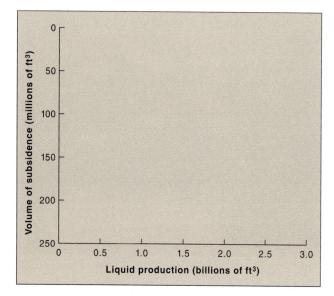

a. On the graph above, plot these two variables against each other and connect the five data points by lines. How do you interpret the shape of the line from 1943 to 1962?

b. Calculate the ratio between the volume of subsidence and liquid production for each of the five time periods. Are the five ratios similar? How might you interpret this result?

6. Because such a high percentage (40%) of America's energy supply comes from oil, the worldwide distribution of this resource is vitally important to the United States. We have only 2% of the world's oil reserves and import about 50% of the oil we use. This single fact is a major influence on America's foreign policy. Without foreign oil the nation would plunge into the worst economic depression in its history and perhaps cease being the world's major economic power.

The data and map on page 141 illustrate some aspects of the problem.

Petroleum Reserves (Billions of Barrels)

World total 1,046 billion barrels

1. Saudi Arabia	262
2. former USSR	119
3. Iraq	100
4. Kuwait	95
5. United Arab Emirates	81
6. Iran	76
7. Venezuela	64
8. Mexico	48
9. Libya	30
10. China	27

Year	American Imports Total	From Middle Eastern Countries
1982	4,298	852
1983	4,312	630
1984	4,715	817
1985	4,286	470
1986	5,439	1,160
1987	5,914	1,272
1988	6,587	1,837
1989	7,202	2,128
1990	7,161	2,243
1991	6,626	2,057
1992	6,938	1,972
1993	7,618	1,995
1994	7,986	1,968

a. There are more than 40 countries that produce significant amounts of oil. What percentage of the world's oil reserves are owned by the top 10 countries?

b. What percentage of the world's oil reserves are located in Saudi Arabia? In the Middle East?

c. Which three countries among the top ten are politically unfriendly toward the United States? What percentage of the world's oil reserves do they own?

d. Locate on the map of the Persian Gulf area the following: Bahrain, Kuwait, Iran, Iraq, Qatar, Saudi Arabia, United Arab Emirates, Strait of Hormuz.

e. Why are the industrialized nations of the world, such as the U.S., Japan, and those of western Europe, so concerned about who controls the Strait of Hormuz?

f. Plot on the graph (p. 141, bottom) the amount of oil imported by the United States since 1982 and the percentage of it that has come from the Middle East.

g. In 1990 Iraq invaded Kuwait and had plans to take over Saudi Arabia as well. The United States went to

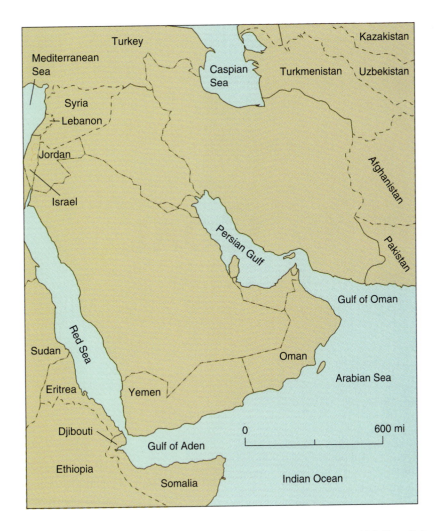

war to prevent this. Do you believe this was justified? Explain your view.

h. The United States is deeply involved in trying to settle the current political dispute between Israel and its Middle Eastern neighbors. How important a factor do you believe America's need for petroleum is in regard to its concern?

Further Reading/References

Cole, H. A. (ed.), 1975. *Petroleum and the Continental Shelf of North-west Europe,* v. 2: *Environmental Protection.* New York, John Wiley & Sons, 126 pp.

Morris, J., House, R., and McCann-Baker, A., 1985. *Practical Petroleum Geology.* Austin, Petroleum Extension Service, University of Texas at Austin, 234 pp.

Selley, R. C., 1985. *Elements of Petroleum Geology.* New York, W. H. Freeman, 449 pp.

Waltham, A. C., 1989. *Ground Subsidence.* New York, Chapman & Hall, 202 pp.

Wardley-Smith, J., 1983. *The Control of Oil Pollution.* London, Graham & Trotman, 285 pp.

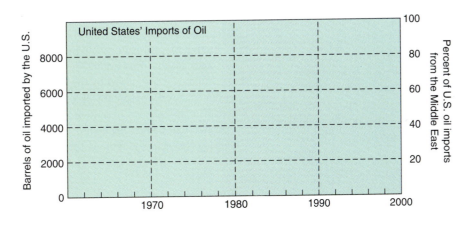

19

COAL

Coal ranks behind oil and gas as a supplier of energy to the American public, accounting for about 25% of our current needs. This percentage has doubled over the past 20 years, despite a 20% growth in total energy consumed. The main reason for the increase is that this fossil fuel is so abundant. About one-fifth of the world's coal reserves are located in the United States (Figure 19.1)—sufficient coal to meet our needs for several hundred years.

Coal deposits are much easier to locate than petroleum and natural-gas deposits because large coal deposits almost always form in swamps. Because coal is a solid, the search is straightforward. If we can find ancient swamp environments in the geologic record, we stand a good chance of finding commercially valuable coal reserves (Figure 19.2). The reason for the association between swamps and coal is that coal forms mostly from partially decomposed remains of land plants. The most luxuriant plant growth occurs in swamps, humid areas where the groundwater table intersects relatively flat ground surfaces that have few through-flowing streams. Many modern coal-forming areas exist on the coastal plain of the eastern United States, from the Carolinas southward into Florida (e.g., Okefenokee Swamp in Georgia). The economically valuable coal reserves found in ancient rocks did not start forming until about 400 million years ago (Devonian Period), when large, multicellular plants first colonized the land surface.

SULFUR CONTENT

As coal forms, the partially decomposed remains of plants pass through several distinctive stages during their transformation to a black, layered, combustible rock (Table 19.1). As the dead plants "mature" toward coal at increasing burial depths and, therefore, at increasing temperature and pressure, their heat output when burned rises, as does the percentage of carbon they contain. Thus, a high rank of coal gives off more heat and leaves less ash residue, which can contain toxic heavy metals. These characteristics are, of course, of interest to any homeowners who burn coal. Most coal, however, is used by industries and electric utilities, for whom sulfur content is even more important. The sulfur is emitted from the smokestack as sulfur dioxide gas (SO_2), which dissolves in atmospheric moisture to become sulfuric acid, a major constituent of acid rain:

$$2SO_2 + O_2 + 2H_2O \rightarrow 2H_2SO_4$$

Plants, the parent material of coal, average less than 1% sulfur, but this percentage can increase greatly during the early stages of coal formation. The enrichment occurs when the swamp is transgressed by the sea during a time of either rising sea level or coastal sinking. Seawater contains abundant sulfur in the form of sulfate ions (2,650 ppm). As the seawater percolates downward through the plant/peat layers, bacterial activity chemically reduces the sulfate to sulfur

Figure 19.1

Coal fields of the United States.

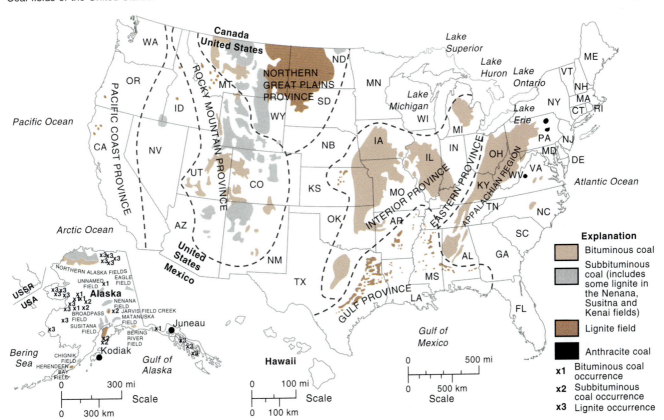

ions (S^{-2}). The ions then combine with available reduced iron (Fe^{+2}) to precipitate as pyrite (FeS_2). Most of the sulfur in high-sulfur coals occurs as microscopic-size pyrite grains, not as organic sulfur. The sulfur content of U.S. coals ranges from 0.2 to 7%, averaging 1 to 2%.

Because of its origin, sulfur content is unrelated to coal rank (Figure 19.3). Western United States reserves contain most of our lower-grade combustible rock, lignite and subbituminous coal, and also contain the highest proportion of low-sulfur coal. But reserves in Appalachia and the Interior Province, all of which contain almost entirely bituminous coal, vary greatly in their proportions of low- and high-sulfur types. This variation has considerable economic and environmental significance. The densely populated East Coast is closest to Appalachian coal, 40% of which has a high sulfur content and must either be treated before burning or "scrubbed" in the exhaust stacks after burning to prevent sulfur emissions. Both treatments increase the cost of the coal but may still be more economical than shipping low-sulfur coal long distances.

COAL MINING AND THE ENVIRONMENT

As a rule of thumb, coal beds to depths of 200 ft are considered surface-minable *(strip mines)*, while beds 200 to 1,000 ft deep are considered underground-minable. Actual mining decisions also depend on the thickness of the coal seam, the quality and rank of the coal, the difficulty of removing overlying rocks, transportation costs, and the prevailing price of the coal. About two thirds of the coal mined in the United States is strip-mined. Coal mining, particularly strip-mining, creates both topographic and esthetic problems. Blasting and use of bulldozers and other equipment removes vegetation and fragments the overburden, greatly increasing both the erosion rate and the sediment load of nearby streams. The sediment can fill streambeds and increase the frequency of flooding. Such problems can be alleviated by rapid land reclamation, now required by federal law. The distressed land must be restored as closely as possible to its premining condition. Smoothing out artificially chopped-up topography and replanting vegetation are costly steps necessary to prevent major environmental damage and esthetic deterioration.

Underground coal mining also creates environmental problems. Acid mine drainage occurs in many mining localities. In areas of older abandoned mines (Figure 19.4), surface structures sometimes collapse into unsupported, shallow underground mines. With increasing age, supporting structures within mines decay, while subsurface waters

Figure 19.2

Coal strip mine near Gillette, Wyoming. The upper half of a 90-foot-thick bed of subbituminous coal is shown here, overlain by 10–60 feet of overburden.

TABLE 19.1

Some Important Characteristics of Organic Material during the Transformation from Living Plants to Coal, Based on Analyses of United States Coal Resources

Substance	Carbon % (excluding moisture)	Moisture + Volatile Matter (%)	Btu per Pound
Living plant	50	—	—
Peat	50–60	>75	—
Lignite	60–70	75	< 9,000
Subbituminous coal	70–78	70	9,000–12,200
Bituminous coal	78–90	60	12,200–15,500
Anthracite coal	92–98	8	~15,000

Figure 19.3

Estimates of 1987 recoverable coal reserves by region and sulfur content. *Source: Energy Information Administration, 1989, p. ix.*

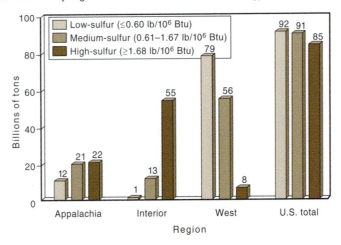

Figure 19.4

Modes of ground subsidence above old, abandoned coal mines. *R. W. Bruhn and others, 1978, Subsidence over the mined-out coal: American Society of Civil Engineers Spring Convention, Pittsburgh, ASCE Preprint 3293, p. 26–55.*

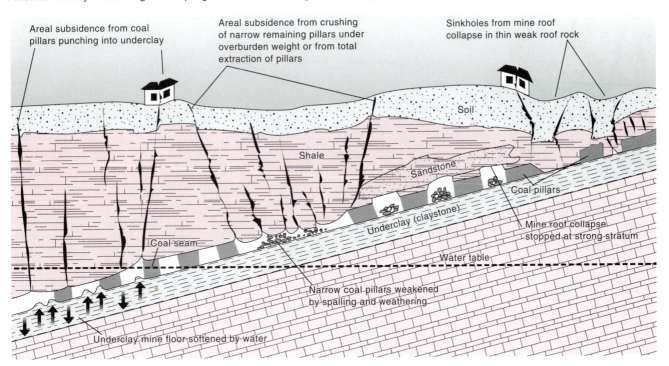

weaken the rocks through weathering. The subsidence typically occurs as pits or trenches that can be deeper than the thickness of the coal mined. A related and perhaps even more serious problem in old, abandoned mines is spontaneous combustion of the remaining coal. In the mid-1980s the U.S. Bureau of Mines estimated that more than 250 uncontrolled mine fires were burning in 17 states (Figure 19.5). Such fires can release noxious sulfurous fumes and increase the frequency of surface-structure collapse. Some fires in Pennsylvania have been burning for 25 years, travelling several kilometers during that time. Periodically, both people and property have fallen into newly collapsing pits

at the surface. Repeated attempts to extinguish the fires have failed. If you live in a coal-mining area, is your home insured against such a disaster?

All the damage coal mining causes to property and the natural environment is minor, however, compared to its damage to the miners themselves. Miners, particularly those who work underground, inhale large amounts of coal dust. The dust particles are too small to see, but they have disastrous effects on the lungs, causing chronic bronchitis, emphysema, and, in severe cases, death. These effects are called black lung disease. The accumulation and retention of coal dust in miners' lungs is directly correlated with

Figure 19.5
A mine fire.

their years of underground exposure (Figure 19.6). Various government studies have found significant coal dust in the lungs of 10 to 30% of current miners and 15 to 75% of ex-miners. Ex-miners have higher percentages, partly from spending longer times in underground mines and partly from the earlier lack of protective measures in the mines. Current measures reduce dust in the mines but cannot eliminate it. Coal mining remains a dangerous occupation.

Problems

1. The Fruitland Formation in northwestern New Mexico and southwestern Colorado is a Cretaceous clastic unit, about 75×10^6 years old. It is about 300 ft thick and consists mostly of interbedded sandstone, siltstone, and shale, but also contains economically important coal beds in its lower part. Geologic studies have established that when the coals were deposited, the land area was generally to the southwest, while the sea was to the northeast. Figure 19.7 shows the total coal thickness in the Fruitland Formation.

   a. Draw a line through the axis of thickest coal deposits on the map.

   b. What does the orientation of this line tell you about the orientation of the shoreline during the time of coal deposition?

   c. Do you think the rate of sinking was the same throughout the entire depositional basin? Why or why not?

   d. Most of the coal occurs in the lower one third of the Fruitland Formation. What does this tell you about drainage conditions during the depositional time of the lower versus the upper Fruitland Formation?

   e. Make a numerical estimate of the volume of coal present in the Fruitland Formation.

2. List as many environmental hazards as you can that can result from mining and burning coal.

3. When living organisms die, their tissues normally decay to carbon dioxide plus water, leaving no residue. Why does this not occur in the environment in which most coal forms? (Hint: swamps are stagnant environments.)

Figure 19.6

Relationship between years worked in underground coal mines and incidence of miners' pneumoconiosis. *Archives of Environmental Health, Volume 27, Page 182, 1973. Reprinted with permission of the Helen Dwight Reed Educational Foundation. Published by Heldref Publications, 1319 Eighteenth St., N.W., Washington, D.C. 20036–1802. Copyright © 1978.*

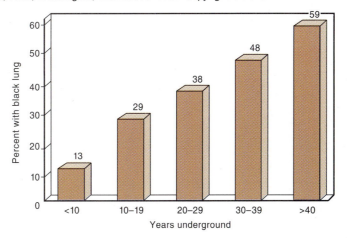

4. Considering the requirements for coal formation from dead plant material, examine a geologic map of the United States to explain why all lignite deposits are located in the Montana–North Dakota area rather than in, for example, the Appalachian region.

5. The amount of heat produced when a hydrocarbon such as coal or oil is burned can be determined using the equation

$$Q = 8400 \text{ C} + 27{,}765 \text{ H} + 1500 \text{ N} + 2500 \text{ S} + 2650 \text{ O}$$

where Q is heat (calories/gram) and C, H, N, S, and O are the weight percents of these elements in the coal or oil.

The table below shows the average compositions of peat and lignite (pre-coal materials), bituminous coal, anthracite coal, and petroleum.

Material	%C	%H	%N	%S	%O
peat	50	0.5	1.9	3.6	44
lignite	70	0.4	1.5	3.1	25
bituminous coal	85	0.5	1.7	2.8	10
anthracite coal	94	0.2	1.1	1.7	3
petroleum	86	12.0	0.5	1.0	0.5

a. Calculate the relative energy contents of these five sources of heat energy.
b. What is the main factor that causes these differences in heat content?
c. If the price per pound were the same for these five materials, which would you buy to heat your home?

6. The Carboniferous Period of earth history (355–290 million years ago) is so named because a major part of the earth's coal reserves are located in rocks of this age. Considering the depositional setting in which coal

precursors live, what might you infer about the latitudinal positions of the earth's large landmasses during this time period?

7. A large percentage of this country's coal reserves are located more than 200 ft underground and cannot be strip-mined. The nation needs access to this coal and will need it even more in upcoming decades. But because of the dangers of underground mining, these resources are underutilized. Suggest steps that might be taken, either by private industry or by local, state, and federal governments, to alleviate this problem.

8. In Pennsylvania some neighborhoods are settling and houses have collapsed because of underground coal mining. Suppose the coal company is still operating in such an area. What should be done? Are the homeowners entitled to financial relief? From whom? What if the coal company is out of business? What then?

Further Reading/References

Chadwick, M. J., Highton, N. H., and Lindman, N., 1987. *Environmental Impacts of Coal Mining and Utilization.* New York, Pergamon, 332 pp.

Corcoran, Elizabeth, 1991. Cleaning up coal. *Scientific American,* May, p. 106–16.

Kottlowski, F. E., Cross, A. T., and Meyerhoff, A. A., 1978. *Coal Resources of the Americas.* Geological Society of America Special Paper No. 179, 90 pp.

Kroll-Smith, J.S., and Couch, S.R., 1990. *The Real Disaster Is Above Ground: A Mine Fire and Social Conflict.* University of Kentucky Press, Lexington, 210 pp.

Lee, F. T., and Abel, J. F., Jr., 1983. *Subsidence from Underground Mining: Environmental Analysis and Planning Considerations.* U.S. Geological Survey Circular 876, 28 pp.

Figure 19.7

Map showing thickness of coal beds in the Fruitland Formation. *Source: U.S. Geological Survey.*

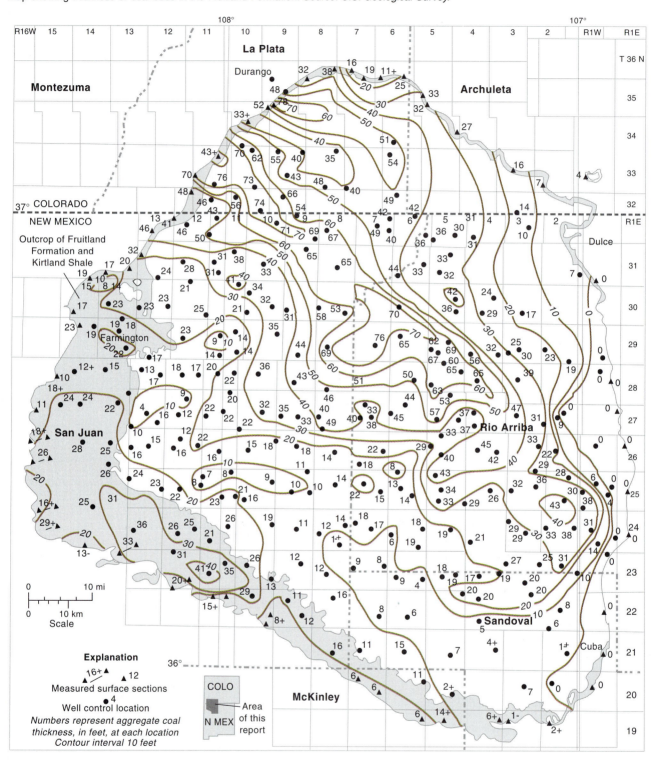

E X E R C I S E 20

AIR POLLUTION

Until the last 25 years or so, no one except meteorologists thought much about the air we breathe. Its composition and purity were taken for granted. Since that idyllic time, however, we have become aware that our air is seriously polluted, not everywhere to the same degree, but polluted to some degree everywhere. The types of pollutants are varied. Some are gases such as carbon dioxide, others are liquids such as *acid rain,* and still others are solids such as *soot.* Where do these materials come from, how do they get into the air, and what can be done about them?

COMPOSITION OF THE ATMOSPHERE

Our air is composed of about 78% nitrogen and 21% oxygen, with trace amounts of other gases such as argon (0.9%), carbon dioxide (0.035%), and even smaller amounts of eight other gases. In addition, air contains water vapor that we sense as humidity and see as rain or snow. Human activities such as driving cars, operating coal-fired power plants, and producing propellants for cans of hair spray have not only increased the amounts of some of the trace gases (which total less than 0.1% of the air) but have added some new things. Did you ever wonder what happens to the tiny bits of rubber that wear from your tires as you drive? Check your lungs.

Most of the noxious pollutants we have added to the air result from burning coal and oil, commonly as the refined product called gasoline. The contaminants that are moni-

tored regularly by urban air-quality agencies are sulfur dioxide, particulate matter, nitrogen oxides, carbon monoxide, ozone, and lead. No major city passes muster for all six categories of contaminant.

Burning coal in power plants to generate electricity produces 80–85% of America's sulfur dioxide emissions. Most of the rest comes from petroleum refining and smelting of sulfide ores. Sulfur dioxide causes lung damage.

Particulate matter consists of solid and liquid particles suspended in the air. These particles are mostly soil and soot. The largest soot generators are the smokestacks of coal-fired power plants. Particles smaller that 2.5 micrometers are the most damaging because they pass through your body's defenses and lodge in your lungs (Figure 20.1).

Nitrogen oxides are produced by the burning of hydrocarbon fuels, especially in power plants and motor vehicles. The harmful effects are the same as those of sulfur dioxide.

Carbon monoxide interferes with the ability of your blood to carry oxygen to your body. Unborn children and those with heart conditions are particularly at risk. The earth's major source of carbon monoxide is the combustion of gasoline in motor vehicles. A catalytic converter in your car engine converts the carbon monoxide to carbon dioxide, better for your health, but increased carbon dioxide in the air may be causing global warming.

Ground-level ozone is another lung irritant, known popularly as *smog.* It forms from a reaction between two things

Figure 20.1

Schematic illustration of the respiratory tract and its anatomic subdivisions. Indicated on the left are the estimated deposition sites for indicated particle sizes. For example, particles of greater than 10 μ in diam are rarely encountered beyond the larynx. *Kluwer Academic Publishers Introduction to the Scientific Study of Atmospheric Pollution, 1971, B. M. McCormac, Copyright © 1971, with kind permission from Kluwer Academic Publishers.*

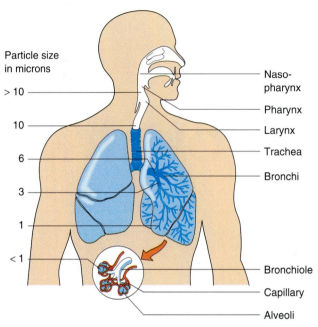

that come from car exhausts, nitrogen oxides and hydrocarbon fumes from incompletely burned gasoline. The Environmental Protection Agency estimates that half the American population is routinely exposed to excessive amounts of ozone. Los Angeles has by far the nation's worst ozone problem, violating the EPA's standard on about 30% of the days each year.

Lead in the air is nearing elimination in the United States, thanks to the phasing out of lead as an additive in gasoline. Many other industrialized countries are also phasing out leaded gas. But lead is still widely used in gasoline in poorer nations, where the number of cars is increasing rapidly.

> When you can't breathe, nothing else matters.
> *American Lung Association*

GLOBAL WARMING

Very large changes in the earth's climate occur without human intervention. Fifty thousand years ago much of North America was covered by a thick sheet of ice, which receded from the United States only 10,000 years ago. It has not yet receded from Greenland. Less extreme climatic variations have happened during the past few thousand years, in about 6000 B.C., 1000 B.C., and from 1450 to 1850, a period known to students of climate as the Little Ice Age. The causes of these climatic changes are uncertain but may relate to changes in solar radiation, sunspots, wobbles in the earth's rotation, or other, still unrecognized causes.

Then there is the *greenhouse effect.* More than 80% of solar radiation is concentrated in the wavelengths our eyes translate as visible light and at the slightly longer wavelengths our skin identifies as heat (Figure 20.2). These wavelengths, ranging from 0.4 to 1.5 micrometers, pass through the atmosphere without being absorbed by the gases that make up the air. The sun's rays hit the earth's surface, are absorbed, and are reradiated at wavelengths greater than 4 micrometers. Water vapor, carbon dioxide, and methane in the air absorb these long *infrared* wavelengths so they cannot escape to outer space. This is the greenhouse effect. As a result, the earth's air and surface temperatures are 63° F warmer than they would otherwise be. More than 90% of the greenhouse effect results from absorption by water vapor. Carbon dioxide absorbs most of the rest.

Carbon dioxide is released by human activities in enormous amounts, largely by the burning of oil and coal. Over the last 150 years, the amount of CO_2 in the air has increased from 280 ppm (0.028%) to 360 ppm (0.036%), an increase of more than 25%. Methane has increased from 0.7 ppm to 1.7 ppm, an increase of more than 100% as a result of livestock and rice cultivation. Molecule for molecule, methane is 60 times more effective as a greenhouse gas than carbon dioxide, but there is 200 times more CO_2 in the air.

The only effective way to reduce the amount of carbon dioxide we pump into the air is to stop using oil and coal and convert to nonpolluting and inexhaustible alternatives such as solar power, wind power, and more exotic energy sources such as tidal power and ocean wave power (Exercise 21).

Figure 20.2

Solar radiation: input, atmospheric absorbtion, and reradiation by the earth.

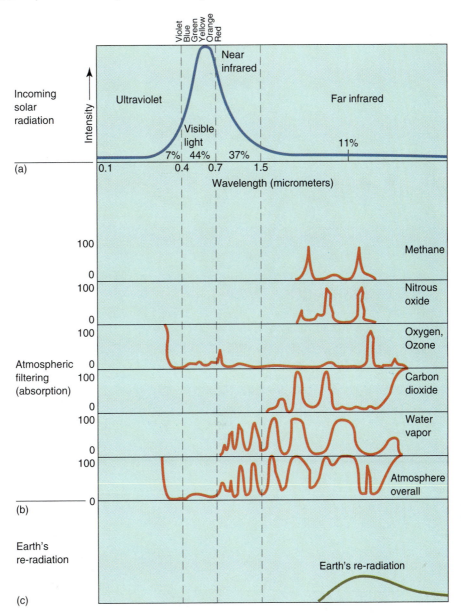

Almost all scientists agree that the earth's average surface temperature has risen by about 1° F over the past 150 years, the period during which the amount of carbon dioxide has been increasing because of human activities. However, it is far from certain that the CO_2 we are pumping into the air is the cause of the change. After all, temperature changes far greater than one degree have occurred repeatedly before humans came onto the scene about a million years ago and also afterward, before oil and coal were being used. The importance of human activities as a cause of climate change is intensely debated among those who study global changes.

OZONE DEPLETION

Essentially all oxygen gas occurs in the air as O_2, two atoms of oxygen bound together as an oxygen molecule. But sometimes three atoms of oxygen are bound together to form O_3, a gas known as ozone. At ground level concentrations of ozone are known as smog, a gas that attacks your eyes and lungs (Figure 20.3). It forms when sunlight stimulates a chemical reaction between gases from your car exhaust.

However, most of the atmosphere's ozone does not exist near the ground. About 90% of it is concentrated in the now-famous ozone layer, 10–22 miles above the ground (Figure 20.4). Here its concentration is about six times that

Figure 20.3

A clear day and a smoggy day in New York, New York.

at sea level. The ozone layer is critical to human (and possibly other) life on earth because it absorbs the very short wavelengths of solar radiation (*ultraviolet* rays). Without the ozone layer there would be sharp increases in eye cataracts, skin cancer, genetic mutations, and a rise in the occurrence of birth defects and damaged immune systems. Increased ultraviolet radiation also disturbs the reproduction

and growth of oceanic plankton, the one-celled organisms at the base of the oceanic food chain. A healthy ozone layer is essential to healthy life on earth.

In 1985, satellite measurements revealed that the ozone layer above Antarctica was thinner than elsewhere. For the next 10 years, the amount of thinning continued to increase and became known as the "ozone hole." The loss of ozone

Figure 20.4

Ozone concentration in the atmosphere as a function of altitudes. *Source:* Ecodecision, *Winter 1996, p. 67.*

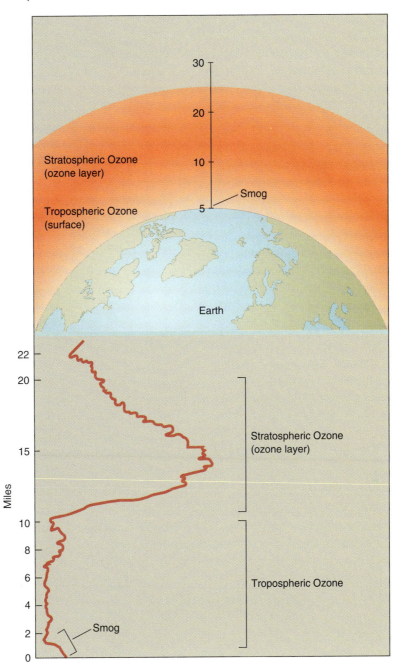

spread over the earth, although not to the extent seen over Antarctica. It is now known that the chemical reaction that destroys the ozone molecules occurs on the surface of ice crystals in the upper atmosphere, so naturally it is most effective above Antarctica.

What chemical reactions caused the ozone loss? The answer was found to be the presence of chlorine atoms. Chlorine reacts with ozone and destroys it. The source of the chlorine was found to be *CFC*s (chlorofluorocarbons), complex molecules invented in the 1920s and widely used as coolants in refrigerators and air conditioners (where it is known as Freon, a Dupont trade name), as cleaning agents for electronic components, as *aerosol* propellants in many spray cans, and in Styrofoam drinking cups. When CFCs rise in the atmosphere and enter the ozone layer, solar radiation breaks them apart, releasing chlorine atoms that attack ozone. Each chlorine atom can destroy 100,000 ozone molecules and a single molecule of a CFC contains lots of chlorine atoms. So a little CFC is extremely effective. CFCs are responsible for 75% of the ozone destruction that has occurred.

The other 25% of ozone loss results from attack by atoms of bromine, an element similar to chlorine. About half of the atmosphere's bromine atoms occur naturally; the other half is released into the environment by human activities, particularly pesticides and the burning of plants—grass and wood (forests). Atom for atom, bromine atoms are 50 times more effective as ozone-destroyers than chlorine, but bromine atoms are 100 times less abundant.

Governments normally move with the slowness of a snail, but the seriousness of the ozone thinning as a threat to human survival spurred immediate action. The production of CFCs was banned worldwide by the year 2000. CFC production peaked in 1988 (only three years after the ozone thinning was discovered) and has decreased by more than 75% as of 1994. Chlorine abundance in the atmosphere peaked in 1994 and has begun a slow decline. Meteorologists estimate it will take about 50 years for the ozone layer to fully recover.

PROTECTION FROM ULTRAVIOLET RADIATION

Ultraviolet radiation penetrates your skin to different depths depending on its wavelength (Figure 20.5). Your skin is particularly transparent to ultraviolet wavelengths of about 315, 366, and 440 micrometers. As shown in Figure 20.2, ozone and oxygen are the only absorbers of these wavelengths, so that depletion of the ozone layer lets more of these wavelengths reach your skin. Skin cancer can result from too much exposure. A sunscreen lotion with a sun protection factor (SPF) of 15 screens out 94% of UVB radiation.

For each 1% decrease in ozone shielding, skin cancer is projected to increase by 2%. At the South Pole concentrations have decreased 70% in recent years, increasing the skin cancer risk (if you lived there) by 140%. The effect on penguins is unknown.

The projected increase in eye cataracts can be minimized by wearing sunglasses treated to absorb ultraviolet rays. Untreated sunglasses can be worse than none at all because your pupils dilate (expand) under dark lenses (in an attempt to capture more light), making it easier for UV rays to damage the retina, the screen at the back of your eye.

Problems

1. Acid rain is formed when sulfur dioxide from power plants and smelters (65%) and nitrogen oxides from car exhausts (35%) react with oxygen in the air to form sulfuric acid. The map on page 155 shows degrees of acidity of rainfall in North America, expressed in pH units.
   a. Contour the data in the eastern half of North America at a contour interval of 0.5 pH units.
   b. Why are pH values so much lower in the eastern part of the continent than in the western half?
   c. Uncontaminated rainfall has a pH of 5.6. How much more acidic than normal is 4.1, the lowest rainfall pH value on the map?
   d. Which lakes in the northeast would have lower pH values because of acid rain, those on granite or those on limestone? Explain why. Answer the same question for granite versus basalt.
   e. What would be the effect on acid rain distribution of increasing the height of smokestacks at factories and power plants that emit sulfurous gases?

2. Litmus, a coloring agent made from certain lichens, is the essential component of litmus paper, which is used to determine the pH of liquids. Litmus paper turns red in acidic solutions and blue in basic ones. Modern refinements of

Figure 20.5

A cross-section of the human skin and underlying tissue, showing the depth to which UV radiation penetrates. The thickness of the epidermis, roughly 50 micrometers (or 1/500 of an inch), is about half the thickness of this page of paper. UV-A radiation is not considered dangerous to humans. *Source:* Consequences, *v. 1, no. 2, 1995, p. 160.*

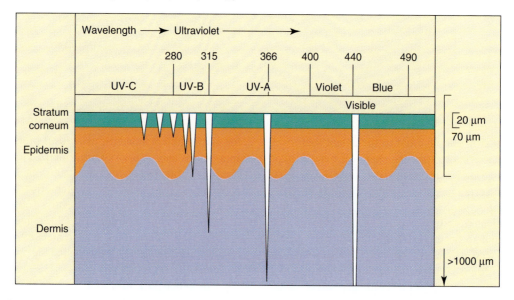

Annual mean value of pH in precipitation, weighted by the amount of precipitation in the United States and Canada. *Source: Environmental Protection Agency.*

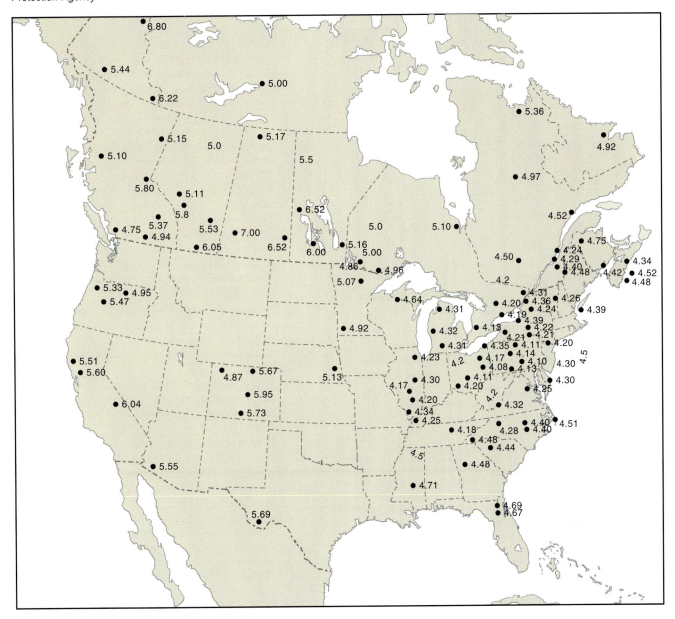

litmus paper that are commercially available for under $20 a package provide a sharp color change for each 0.5 pH unit from pH 0 to pH 14.

Using the litmus paper provided, determine the pH of each of the following.
a. tap water
b. commercial bottled water
c. vinegar
d. tap water as a student blows into it through a straw
e. water containing powdered gypsum
f. water plus baking soda

Explain why the pH is acidic or basic for each measurement.

3. Collect small bottles of water from local rainfall, lakes, and streams and determine their pH values using the litmus paper provided. Explain your results.

4. The 49 data points on the graph on page 156 are from one of many studies that relate the amount of particulates in the air to death in American cities. The sampling procedure was designed to make sure the deaths were not related to smoking, educational level, or age.

a. Does this look to your eyes to be a significant trend? Eyeball your estimate of the best fit straight line for the data.

b. Ask your instructor to show you how to calculate the best fit straight line for the data and to determine the degree of relationship between the two variables. (The hand calculation is tedious and time-consuming. There is a computer program for the calculation.)

c. In this investigation, fine particles were defined as those with diameters smaller than 10 micrometers (0.004 inches). What do you think the scattering of points

Relationship between death and particulates in the air smaller than 10 micrometers. *From the* American Journal of Respiratory and Critical Care Medicine, *1995, vol. 151.*

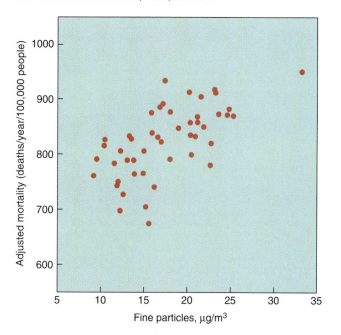

would be if the definition included all particles smaller than 30 micrometers? Suppose only particles with sizes smaller than 2.5 micrometers were included. What do you think the scatter plot would look like then?

d. This study considered only the size of the particulates. Do you believe that differences in the type of particulate would have an effect on these results? Explain why.

5. There is no question that the combustion of gasoline is a major cause of air pollution. The table below shows the relationship between the amount of tax a government puts on gasoline and the amount of gasoline each person in the country uses (in gallons).

Gasoline Tax and Gasoline Use in Some Countries in 1993

Country	Tax ($/gallon)	Use (gallons/person/year)
United States	0.36	400
Canada	0.84	281
Australia	0.96	234
Japan	1.20	91
Germany	1.92	124
Sweden	2.20	158
Italy	2.88	100
Portugal	3.40	59

a. Make an X-Y scatter plot of these data. Does the correlation look better or worse than the one in

problem number 4? Then ask your instructor for the correlation coefficient.

b. What factors other than the amount of gasoline tax might be responsible for the less-than-perfect correlation?

6. Whether the cause is natural variation, human activities, or both, global temperatures are rising. So is sea level. Your instructor has given you two topographic maps, one of an area on the mid-Atlantic coast, the other on the west coast.

a. Using tracing paper and a colored pencil, show where the shoreline would be if global warming caused most of the Antarctic ice sheet to melt and sea level rose 100 feet.

b. Which side of the United States would suffer more from the rise in sea level? Why?

c. Examine a map of the United States and determine which American cities would be flooded. Assuming the flooding is inevitable and will occur gradually over a period of 200 years, what do you think can or should be done about it? Would your answer change if the rise were going to occur within 50 years? Why?

7. A decrease in thickness of the ozone layer in the upper atmosphere will result in an increase in occurrence of skin cancer. Examine the graph on page 158, based on data accumulated before the ozone layer was thinned by human activities.

a. Why is skin cancer more common in Florida and Texas than in other states?

b. Do you believe all races would be affected to the same degree? Explain.

c. What other factors might affect skin cancer rates in the states?

8. The maps and graph on page 158 show the daily and monthly variations in ozone concentration in southeastern Canada. Canadian law sets 82 ppb as the maximum permissable level. Prevailing winds are shown on the map.

a. In which areas is ozone (smog) the worst?

b. The occurrence of smog peaks is greatest during summer months and is least during winter months in Toronto (and in Montreal as well). Can you explain this?

c. Suppose the monthly graph were a daily graph from midnight on the left to midnight the next day on the right. What do you think would be the shape of the graph? Why?

d. The sample location map shows the location of the major industrial cities in the northeast. What and where might you suspect is the source of the smog in Toronto and Montreal? How does this compare with the data for acid rain in question number 1?

e. Do you believe southeastern Canada has a serious smog problem? Why or why not?

Further Reading/References

American Lung Association, 1992. *Health Effects of Ambient Air Pollution.* New York, 63 pp.

Relationship between latitude and skin cancer in the United States.

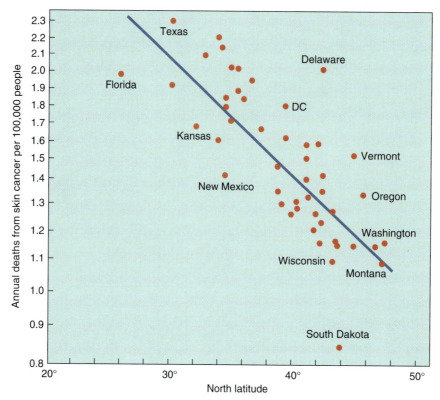

Locations of ozone monitoring sites in eastern Canada. *Source:* Journal of the Air and Waste Management Association, *1994, v. 14, p. 1020.*

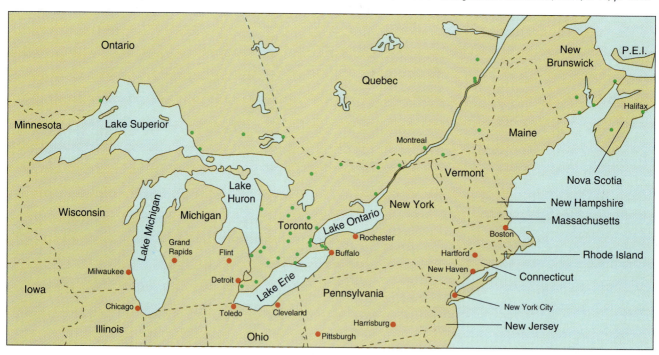

Brink, Susan, and Wu, Corinna, 1996. "Sun struck." *U.S. News & World Report,* June 24, p. 62–67.

Gribbin, John, and Gribbin, Mary, 1996. "The greenhouse effect." *New Scientist* (insert), July 6, 4 pp.

Lee, John, and Manning, Lucy, 1995. "Environmental lung disease." *New Scientist* (insert), Sept. 16, 3 pp.

Mage, David and others, 1996. "Urban air pollution in megacities of the world." *Atmospheric Environment,* v. 30, p. 681–86.

Royal Swedish Academy of Sciences, 1995. "Environmental effects of ozone pollution." *Ambio* (special issue), v. 24 no. 3, May, p. 137–96.

Spatial distribution of average number of days when ozone concentration exceeded 82 parts per billion from 1980 to 1991 in southern Ontario and Quebec. *Source:* Journal of the Air and Waste Management Association, *v. 14, 1994, p. 1024.*

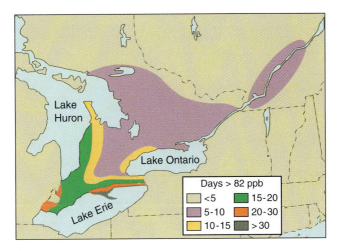

Prevailing wind directions in the northeastern United States. *From W. D. Bischoff, et al.,* Geological Aspects of Acid Rain Deposition, *edited by O. P. Bricker. Copyright © 1984 Butterworth Publishers, Boston, MA. Reprinted by permission.*

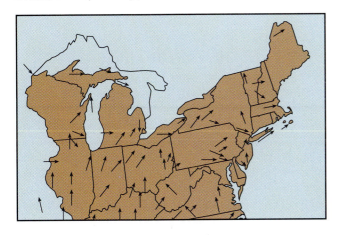

Variation in ozone concentration in Toronto in 1991. *Source:* Journal of the Air and Waste Management Association, *v. 14, 1994, p. 1020.*

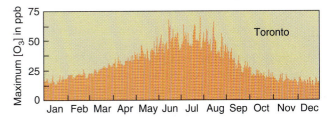

ALTERNATIVE SOURCES OF ENERGY

About two thirds of the energy needed to power American industry is obtained from oil and coal. The percentage is about the same in most other industrialized nations. However, these two fuels are the source of most of America's (and the world's) air pollution problems and, because of oil refineries, a major cause of water pollution as well. Also, oil and coal are nonrenewable, meaning they are being used at a rate many times faster than they are formed by natural processes. Oil and coal take millions of years to form but America's and the world's supplies will be exhausted within 50–200 years; first oil, then coal.

All industrialized countries rely on oil, natural gas, and coal (Figure 21.1). Nuclear energy amounts to less than 10% of the energy used in most industrialized nations. The terrible effects of the steam explosion and partial core meltdown at Chernobyl in Ukraine in 1986 caused many of the industrialized nations to reconsider the use of nuclear energy.

Undeveloped nations rely on burning wood, with hydropower as a secondary source. But burning wood is not a meaningful alternative for nations with the industrial infrastructure of Europe, North America, or Japan. Wood is also nonrenewable at the rate it is now being used, as the destruction of the world's rainforests shows.

Are there renewable, inexhaustible, and nonpolluting sources of energy in amounts large enough to satisfy the world's continually growing demand for energy? If so, what are they?

THE ALTERNATIVES

There are many alternatives to fossil fuels, such as harnessing the energy of ocean waves and tides and the earth's natural heat below the surface. However, the best and, at present, most available alternative sources of energy to power the world's economies are wind power and solar power. These sources are inexhaustible, do not pollute, are available in every country, and are now available at prices that are competitive with oil and coal, about $0.05–0.06 per kilowatt-hour of energy produced. A study conducted in 1994 found that 73% of the American public favors cutting the use of nonrenewable fossil fuels and nuclear energy. Forty-two percent believed there should be increased emphasis on the development of renewable sources of energy. Energy efficiency and conservation were major concerns of those polled. Despite this clear public support for renewable and nonpolluting energy, the federal budget for renewable energy technologies was cut 29% from 1995 to 1996.

ENERGY FROM THE WIND

Wind farms are appearing at many places in the United States, particularly in California and in the midwest (Figure 21.2). The reason is easy to see (Figure 21.3); it's pretty windy there. It turns out that the power available from wind increases as the cube of the wind speed, so a doubling of

Figure 21.1

Global consumption of energy (million tons of oil equivalent). Underdeveloped nations (olive color) rely overwhelmingly on biomass as an energy source.

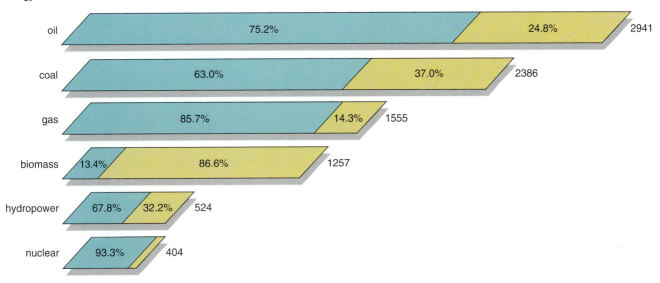

oil — 75.2% | 24.8% | 2941

coal — 63.0% | 37.0% | 2386

gas — 85.7% | 14.3% | 1555

biomass — 13.4% | 86.6% | 1257

hydropower — 67.8% | 32.2% | 524

nuclear — 93.3% | 404

Figure 21.2

Modern windmill array for power generation. Tehachapi Pass, California. © Wm. C. Brown Communications, Inc/Douglas Sherman, photographer.

wind speed increases the power available by eight times. A tripling of wind speed results in an increase of 27 times. The destructive power of hurricanes and tornadoes shows what strong winds can do when not harnessed.

The largest wind farms are in California, where wind produces enough electricity for the residential load of a city the size of San Francisco. Wind turbines in California are concentrated in mountain passes where the wind is strong and steady. Altamont Pass east of San Francisco has 7,500 turbines; another large group is in Tehachapi Pass east of Los Angeles. There are about 17,000 wind turbines operating in California, and the state currently produces one third of the world's wind-generated electricity. However, the percentage is dropping rapidly as the technology spurts worldwide. Just 10 years ago California had 85% of the world's capacity. The world's wind-generating capacity has increased from only 10 megawatts in 1980 to 1,020 mW in 1985, 1,930 in 1990, 4,700 in 1995, and jumped to 5,800 mW in 1996. Europe now generates half the world's wind power.

Until the 1990s, ideas for tapping wind energy centered around sites on land. But wind speeds are several times greater offshore than onshore and also are steadier because there are no topographic barriers to wind movement (Figure 21.3). Danish companies built the world's first offshore wind farm in 1991, a mile offshore in water 10–20 feet deep.

ENERGY FROM THE SUN

The most intense solar radiation falls at low latitudes. The equator receives 2.5 times the radiation received at the poles.

Figure 21.3

Wind power available at 150-foot height. *Reprinted with permission of Macmillan Library Reference USA, a Simon & Schuster Macmillan Company, from ATLAS OF UNITED STATES ENVIRONMENTAL ISSUES by Robert J. Mason and Mark T. Mattson. Copyright © 1990 by Macmillan Publishing.*

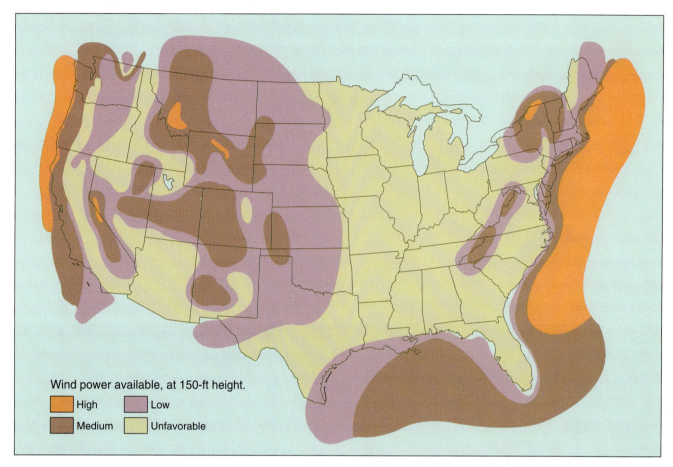

Wind power available, at 150-ft height.
- High
- Low
- Medium
- Unfavorable

Most of the earth's poorest nations are located in low latitudes (for example, India and most of Africa) and contain few fossil fuels; solar energy can be an important factor in their future development. The potential for solar power in the United States is shown in Figure 21.4. In most of the country, solar power could be used for more than 70% of the year and is usable practically everywhere for at least half the year. Figure 21.4 was drawn in 1978. The amount of solar radiation that can be captured by solar cells has increased greatly since then, so that the same map drawn today would show an increase in the size of the most suitable areas.

SOLAR CELLS

Solar cells, also called photovoltaic cells, convert light energy directly into electricity. Most solar cells are thin wafers of the element silicon, to which small amounts of gallium and cadmium have been added (Figure 21.5). When hit by the sun's rays, the wafers give off electrons that are carried along a wire as electricity. Large groups of cells are wired together to increase the amount of electric power generated.

Current solar cells convert 15–20% of light energy into electrical energy, and experimental solar cells have reached efficiencies of 30%. As the efficiency improves, solar power generation becomes more economical and, therefore, practical at higher latitudes. At 20% efficiency, an area about 76 miles square packed with solar cells could supply all of America's present demand for electricity. This area of 5,800 square miles is slightly smaller than the combined areas of Connecticut and Rhode Island.

However, silicon-based solar panels may soon be obsolete. In 1994 researchers developed an organic polymer film that generates electricity from sunlight and could provide it at 1% of the cost of solar cells. The film can be rolled like wallpaper for easy transport, and when laid flat in sunlight it converts solar energy into electrical energy.

Unfortunately, the United States seems less interested in solar power than other countries. Between 1981 and 1990, the U.S. share of the worldwide solar cell market fell from 75% to 32%, while Japan's share grew from 15% to 32%. The world's largest producer of solar cells is a foreign company, and in 1995 it expanded its production capacity by 50%. What do you think the relative degree of energy independence of Japan and the United States is likely to be when the supply of nonrenewable fuels becomes unaffordable?

Figure 21.4

Availability of solar energy during the day in the continental United States. *U.S. Department of Energy and National Wildlife Federation.*

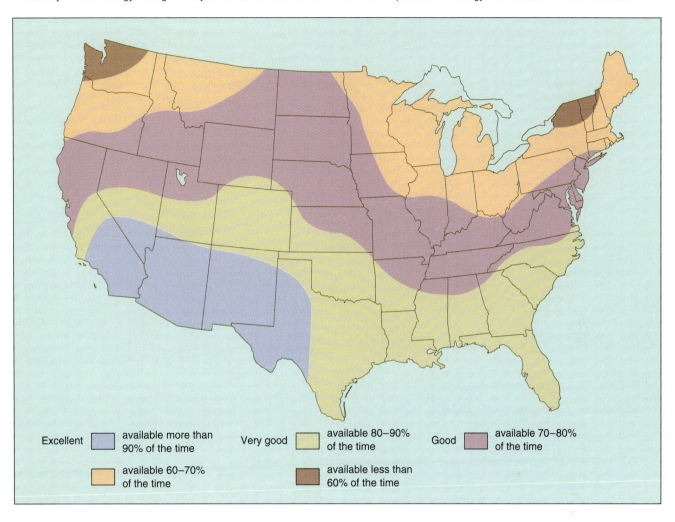

Excellent — available more than 90% of the time

Very good — available 80–90% of the time

Good — available 70–80% of the time

available 60–70% of the time

available less than 60% of the time

Figure 21.5

Solar house in Boulder, Colorado, elevation 5,500 feet, latitude 40 degrees North.

THE BEST ENERGY MIX

No single energy source will ever be the best choice to supply everyone's needs everywhere all the time (Figure 21.6). Solar power will always be cheaper in Florida than in Alaska. Wind power will always be more effective in North Dakota than in Alabama. And there are other alternatives to the polluting fossil fuels we now depend on. Most are in early stages of research and development, so their potential is uncertain. Examples include harnessing the power of ocean tides and waves. An even more exotic possibility is converting the difference in heat energy between shallow and deep ocean waters into commercially usable power. Much effort is also being devoted to the development of fuel cells, devices that combine oxygen (from the air) and hydrogen (produced artificially) to produce electricity. No doubt new ideas for power generation will appear over the next few decades.

Problems

1. What are the reasons that poor countries rely so heavily on biomass as an energy source?

2. Many countries exploit the earth's internal heat to generate electricity (geothermal power).
   a. Where do you think most of them are located in relation to the structure of the earth's crust?
   b. Only four states in the U.S. use geothermal energy to generate electricity (table following). Do the locations of the states make sense in terms of your answer to the first part of this question? How?

Figure 21.6

Areas most suitable for the development of solar power and wind power generation. *Adapted from U.S. Energy Atlas, 1980.*

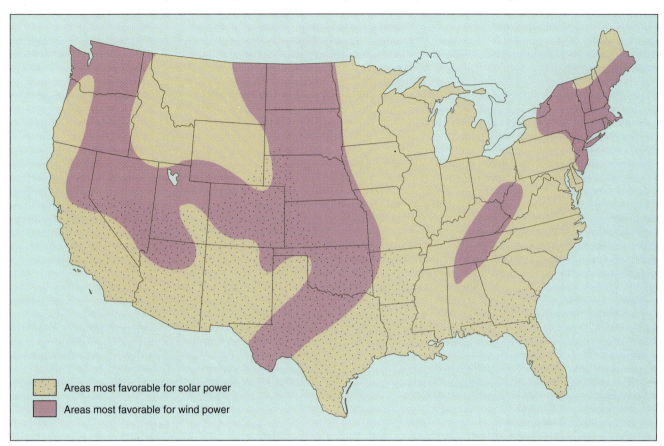

Areas most favorable for solar power

Areas most favorable for wind power

c. What percentage of generating capacity is in California?

d. Examine a geologic map of the U.S. and suggest locations where geothermal energy might be exploited.

Electrical Capacity, mW

California		Nevada	
The Geysers	1,971	Dixie Valley	50
Coso	256	Steamboat Springs	18
Salton Sea	214	Soda Lake	16
East Mesa	102	Beowawe	15
Heber	92	Stillwater	13
Long Valley Caldera	43	Desert Peak	9
Honey Lake	30	San Emidio Desert	3
Amadee Hot Springs	3	Wabuska	2
Hawaii		**Utah**	
Puna	25	Cove Fort	13
		Roosevelt Hot Spring	20

3. What are the geographic characteristics of areas that are suitable for wind farms? Apply your criteria to explain the general pattern of Figure 21.3.

4. Wind speeds vary from zero to perhaps 200 mph in a strong hurricane. Assume an energy of one unit for a wind velocity of 10 mph. Construct a graph showing the change in energy obtained from a wind turbine at wind speeds of 10 mph, 40 mph, 80 mph and 120 mph.

5. Explain why the southwestern U.S. is more favorable for solar power development than the southeastern part of the country at the same latitude (Figure 21.4).

6. What do you think would be the best energy mix for the area where you live? Explain your choice.

Further Reading/References

Anonymous, 1991. *America's Energy Choices.* Cambridge, Massachusetts, The Union of Concerned Scientists, 124 pp.

Davis, G. R., 1990. "Energy for planet Earth." *Scientific American,* March, p. 55–74.

Elliot, D. L., Holladay, C. G., Barchet, W. R. and others, 1987. *Wind Energy Resource Atlas of the United States.* Golden, Colorado, Solar Energy Research Institute, 210 pp.

Kozloff, Keith L., and Dower, Roger C., 1993. *A New Power Base: Renewable Energy Policies for the Nineties and Beyond.* Washington, D.C., World Resources Institute, 196 pp.

Weinberg, Carl J., and Williams, Robert H., 1990. "Energy from the Sun." *Scientific American,* September, p. 146–55.

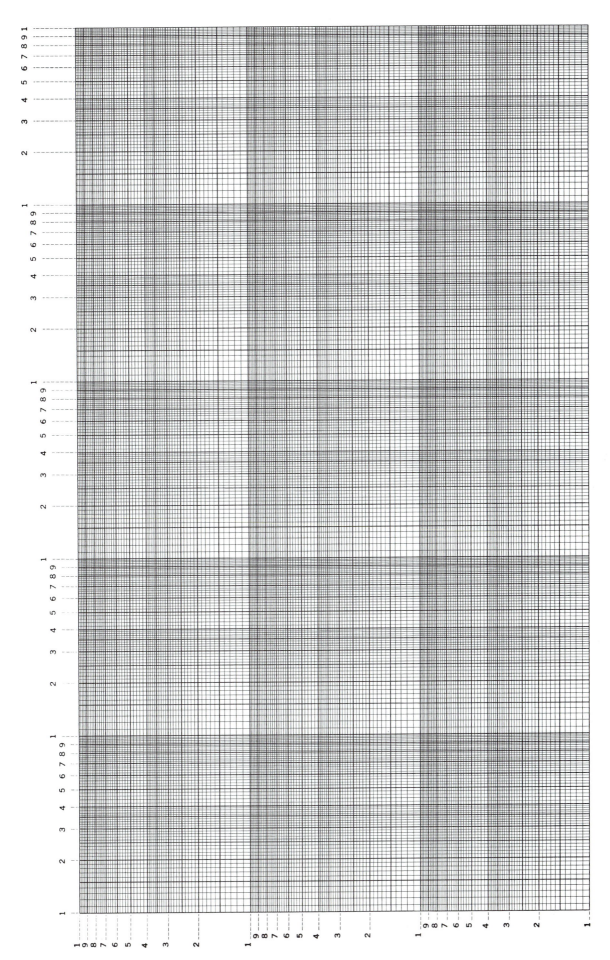

A

Mathematical Conversion Factors

To Convert from	To	Multiply by
Acre–feet	Gallons	3.26×10^5
Acre–foot	Cubic meters	1,233.5
Acres	Square feet	43,560
Barrels oil (bbl)	Cubic feet (ft^3)	5.61
Barrels oil	Gallons (gal)	42
Bars	Pounds/square inch (lb/in.^{2})	14.504
Centimeters (cm)	Inches (in.)	0.394
Centimeters per second (cm/s)	Feet per day (ft/day)	2,835
Centimeters per second	Gallons per day per square foot (gal/day/ft^2)	21,200
Centimeters per second	Meters per day (m/day)	864
Cubic centimeters (cm^3)	Cubic inches (in.^{3})	0.061
Cubic inches (in.^{3})	Cubic centimeters (cm^3)	16.387
Cubic feet (ft^3)	Barrels of oil (bbl)	0.18
Cubic feet	Cubic meters (m^3)	0.028
Cubic feet per second (ft^3/s)	Cubic meters per second (m^3/s)	0.003
Cubic meters	Acre-feet	8.11×10^{-4}
Cubic meters (m^3)	Cubic feet (ft^3)	35.249
Cubic meters per second (m^3/s)	Cubic feet per second (ft^3/s)	353.107
Cubic miles (mi^3)	Cubic kilometers (km^3)	4.167
Cubic kilometers (km^3)	Cubic miles (mi^3)	0.240
Feet (ft)	Meters (m)	0.305
Feet per day (ft/day)	Centimeters per second (cm/s)	3.53×10^{-4}

To Convert from	To	Multiply by
Feet per mile (ft/mi)	Meters per kilometer (m/km)	0.188
Gallons	Acre-feet	3.07×10^{-6}
Gallons per day per square foot (gal/day/ft^2)	Centimeters per second (cm/s)	4.72×10^{-5}
Grams (g)	Ounces (oz)	0.035
Hectares (ha)	Square feet (ft^2)	1.076×10^5
Inches (in.)	Centimeters (cm)	2.540
Kilograms (kg)	Pounds (lb)	2.205
Kilometers (km)	Miles (mi)	0.621
Liters (l)	Quarts (qt)	1.057
Meters (m)	Feet (ft)	3.281
Meters	Yards (yd)	1.094
Meters per day (m/day)	Centimeters per second (cm/s)	0.00116
Meters per kilometer (m/km)	Feet per mile (ft/mi)	5.283
Miles (mi)	Kilometers (km)	1.609
Ounces (oz)	Grams (g)	28.350
Pounds (lb)	Kilograms (kg)	0.454
Quarts (qt)	Cubic centimeters (cm^3)	946.358
Quarts	Liters (l)	0.946
Square centimeters (cm^2)	Square inches (in.^{2})	0.155
Square feet (ft^2)	Square meters (m^2)	0.093
Square feet	Hectares (ha)	0.929×10^{-5}
Square inches (in.^{2})	Square centimeters (cm^2)	6.452
Square kilometers (km^2)	Square miles (mi^2)	0.386
Square meters (m^2)	Square yards (yd^2)	1.196
Square miles (mi^2)	Square kilometers (km^2)	2.589
Square yards (yd^2)	Square meters (m^2)	0.836
Yards (yd)	Meters (m)	0.914
Meters (m)	Kilometers (km)	10^{-3}
Meters	Centimeters (cm)	10^2
Meters	Millimeters (mm)	10^3
Meters	Micrometers (microns; μm)	10^6

B

GEOLOGIC TIME SCALE

Era	Period	Epoch	Start of Interval (Million Years Before Present)	Length of Interval (Million Years)	Percent of Geologic Time
	Quaternary	Holocene	0.01	0.01	0.0002
		Pleistocene	1.8	1.79	0.04
Cenozoic		Pliocene	5	3.2	0.07
		Miocene	23	18	0.39
	Tertiary	Oligocene	34	11	0.24
		Eocene	53	19	0.41
		Paleocene	65	12	0.26
	Cretaceous		135	70	1.52
Mesozoic	Jurassic		205	70	1.52
	Triassic		250	45	0.98
	Permian		290	40	0.87
	Pennsylvanian		320	30	0.65
	Mississippian		355	35	0.76
Paleozoic	Devonian		410	55	1.20
	Silurian		438	28	0.61
	Ordovician		510	72	1.57
	Cambrian		570	60	1.30
Neoproterozoic	Cryogenian		850	280	6.09
	Tonian		1,000	150	3.26

Era	Period	Epoch	Start of Interval (Million Years Before Present)	Length of Interval (Million Years)	Percent of Geologic Time
Mesoproterozoic	Stenian		1,200	200	4.35
	Ectasian		1,400	200	4.35
	Calymmian		1,600	200	4.35
Paleoproterozoic	Statherian		1,800	200	4.35
	Orosirian		2,050	250	5.43
	Rhyacian		2,300	250	5.43
	Siderian		2,500	200	4.35
Archean			4,600	2,100	45.65
Totals				4,600	100.00

C APPENDIX

EARTH SCIENCE INFORMATION CENTERS

The Earth Science Information Centers (ESICs) offer nationwide information and sales service for United States Geological Survey (USGS) maps and earth-science publications. This ESIC network provides information about geologic, hydrologic, topographic, and land-use maps, books, and reports; aerial, satellite, and radar images and related products; earth science and map data in digital form and related applications software; and geodetic data. Write to any ESIC for addresses of the over 60 state ESIC offices.

Any ESIC can fill orders for custom products such as aerial photographs, orthophotoquads, digital cartographic data, and geographic names data.

ESICs can also provide information about earth-science materials from many public and private producers in the United States using automated catalog systems for information retrieval and research services.

For further information contact any ESIC or call 1–800–USA-MAPS.

NATIONAL DISTRIBUTION CENTERS

Lakewood–ESIC
Box 25046, Federal Center, MS 504
Building 25, Rm. 1813
Denver, CO 80225–0046
303–236–5829; FTS 776–5829

Reston–ESIC
507 National Center
Reston, VA 22092
703–648–6045; FTS 959–6045

Washington, D.C.–ESIC
U.S. Department of the Interior
1849 C Street, NW, Rm. 2650
Washington, D.C. 20240
202–208–4047; FTS 268–4047

REGIONAL DISTRIBUTION CENTERS

Anchorage–ESIC
4230 University Drive, Rm. 101
Anchorage, AK 99508–4664
907–786–7011; FTS 868–7011

Anchorage–ESIC
Room G-84
605 West 4th Avenue
Anchorage, AK 99501
907–271–2754; FTS 868–2754

Denver–ESIC
169 Federal Building
1961 Stout Street
Denver, CO 80294
303–844–4169; FTS 564–4169

Los Angeles–ESIC
Federal Building, Rm. 7638
300 North Los Angeles Street
Los Angeles, CA 90012
213–894–2850; FTS 798–2850

Menlo Park–ESIC
Building 3, MS 532, Rm. 3128
345 Middlefield Road
Menlo Park, CA 94025
415–329–4309; FTS 459–4309

Rolla–ESIC
1400 Independence Road
Rolla, MO 65401
314–341–0851; FTS 759–0851

Salt Lake City–ESIC
8105 Federal Building
125 South State Street
Salt Lake City, UT 84138
801–524–5652; FTS 588–5652

San Francisco–ESIC
504 Custom House
555 Battery Street
San Francisco, CA 94111
415–705–1010; FTS 465–1010

Sioux Falls–ESIC
EROS Data Center
Sioux Falls, SD 57198
605–594–6151; FTS 753–7151

Spokane–ESIC
678 U.S. Courthouse
West 920 Riverside Avenue
Spokane, WA 99201
509–353–2524; FTS 439–2524

Stennis Space Center–ESIC
Building 3101
Stennis Space Center, MS 39529
601–688–3544; FTS 494–3544

D APPENDIX

JOURNALS WITH ENVIRONMENTAL CONCERNS

A very large number of professional journals publish articles that deal with environmental problems, and the number is increasing rapidly as concern over environmental problems mounts. Some journals have catchy names such as *Sludge* or *Garbage,* but most of the titles are more drab. The list below includes many frequently cited monthly, bimonthly, or quarterly periodicals; references in the articles they contain will quickly lead you to other important journals. In addition to the periodicals listed, organizations such as the U.S. Geological Survey, the Environmental Protection Agency, and various arms of state governments issue at irregular intervals publications that discuss environmental problems.

Ambio
Archives of Environmental Health
Coastal Management
Consequences
E: The Environmental Magazine
Earth
Environment
Environmental Geochemistry and Health

Environmental Geology
Environmental Impact Assessment Bulletin
Environmental Protection
Environmental Science and Technology
Greenpeace Magazine
Journal of the Air & Waste Management Association
Journal of Coastal Research
Journal of Contaminant Hydrology
Journal of Environmental Economics and Management
Journal of Environmental Management
Journal of Environmental Quality
Journal of Environmental Science and Health
Journal of Environmental Systems
Journal of Soil Contamination
National Geographic
New Scientist
Resources, Conservation, and Recycling
Scientific American
Water Research
Water Resources Bulletin
Water Resources Research
World Watch

GLOSSARY

accessory mineral   mineral whose presence in a rock is not essential to its classification; minor mineral

acid mine drainage   acid waters from a mine because of the conversion of sulfide in minerals to sulfate in water

acid rain   rain more acidic than about 5.6 on the pH scale

active volcano   volcano that has erupted within the past 10,000 years

aerosol   mechanical mixture of a gas and a colloid, either solid or liquid

agglomerate   rock composed of coarse angular volcanic sediment

A-horizon   uppermost soil sediment horizon; also called topsoil

amorphous   noncrystalline

anticline   convex upward fold in a rock layer

aquiclude   impermeable rock overlying an aquifer

aquifer   rock that transmits water in large amounts

artesian well   well in which the water rises above its host rock

atom   smallest unit of an element, composed of a nucleus and electrons

bedding   layering of sedimentary rocks

bed load   sediment carried along the stream bottom

bench mark   an elevation determined accurately by the U.S. Geological Survey

B-horizon   soil horizon below the A-horizon, the subsoil and zone of accumulation

breakwater   offshore artificial structure that breaks the force of the waves against the shore

cement   mineral matter that binds loose sediment into a rock

cementation   process of turning sediment into rock; same as lithification

CFCs   chlorofluorocarbons; complex artificial compounds made of carbon, fluorine, and chlorine

channelization   artificial modification of a stream channel so it can hold more water

chert   rock composed of microcrystalline quartz precipitated from solution

conchoidal fracture   fracture that gives a smoothly curved surface; characteristic of noncrystalline solids

C-horizon   soil horizon below the B-horizon consisting of broken bedrock

clay mineral   mineral group characterized by a sheeted crystal structure and composed largely of silicon and aluminum

cleavage   breaking of a mineral along certain planes because of weak binding across the planes

colloid   material with particle size smaller than 2 micrometers, about 10^{-4} inches

color   wavelengths of light that the eye receives and can process

competence   the size of the largest particle a stream can transport

conglomerate   fragmental rock composed of particles greater than 2 mm in diameter

contact   the surface that separates two types of rock

contaminant   presence in water of a substance in larger amount than occurs naturally

continental divide   line on a map separating water that drains to the Pacific Ocean from water that drains to the Atlantic Ocean

contour interval   difference in value between two adjacent contours

cross-cutting relationship   rocks that cut across other rocks must be younger than the rocks they cut

cross-section   drawing or outcrop of a vertical plane that shows the arrangement of rocks beneath the surface

crust   outermost shell of the earth, a few tens of miles thick

dip   angle of slope of a rock layer into the ground, as measured at 90° to the strike

dormant volcano   volcano that is fresh-looking but has not erupted within the past 10,000 years

drainage basin   the area from which a stream obtains its water

earthquake   sudden trembling in the earth caused by rock rupture

electron   negatively charged particle in an atom

element   atom with a specific number of protons

enzyme   organic substances in living organisms that cause changes in other substances by catalytic action

ephemeral stream   stream that flows briefly only in response to precipitation nearby; intermittent stream

epicenter   point on the surface directly above an earthquake focus

equigranular   rock in which all crystals of major minerals are about the same size

equilibrium   condition when there is no drive for a chemical reaction to occur

evaporite   mineral or deposit composed of very soluble minerals

extinct volcano   volcano that is significantly eroded and has not erupted for more than 10,000 years

fault   break in the earth's crust along which movement has occurred parallel to the breakage surface

fissility   breakage of a mudrock along closely spaced parallel planes because of parallel orientation of clay minerals

floodplain   strip of sediment-covered land adjacent to a stream channel; formed naturally by the stream when it overflowed its banks

floodway   area on the sides of a river that is covered by water during floods of a certain frequency

fold   bend in a rock layer perpendicular to the layering

foliation   planar arrangement of minerals in a metamorphic rock

formation   mappable rock unit

fossil fuel   energy source that formed millions of years ago at a rate much slower than it is used by humans; coal, natural gas, and petroleum

geographic pole   one end of an imaginary line through the earth's center and around which the earth rotates

geologic map   map that shows the distribution, lithology, and age relationships of rocks in an area

glass   noncrystalline solid

gneiss   metamorphic rock having banding formed by layers of different mineral composition

greenhouse effect   heating of the earth's air and surface caused by absorption of the earth's infrared radiation by atmospheric gases

groin   artificial structure perpendicular to the shoreline; designed to stop longshore sediment movement

groundwater   subsurface water below the water table

hardness   comparative resistance to scratching of one mineral by another

hazardous   materials that are either toxic, flammable, corrosive, explosive, or radioactive

humus   dark-colored, decomposed, unidentifiable organic matter in soil

hydrocarbon   chemical compound composed mostly of carbon and hydrogen

hydrograph   graph that shows stream discharge at a location as a function of time

hydrostatic   pressure that is equal in all directions around an object; pressure in a liquid

igneous   rock that solidified from a molten condition

illite   one of three types of clay minerals, rich in potassium

inclusion   fragment of an older rock in an unrelated igneous rock

infrared radiation   radiation at wavelengths between 0.7 micrometers and about 1 mm

inorganic   not formed by living organisms; lacking organic carbon

intensity   scale of relative damage produced by an earthquake

intermittent stream   ephemeral stream

ironstone   sedimentary iron ore composed of sand-sized balls of hematite

isohyet   contour line connecting points of equal rainfall

isopach   line drawn through points of equal thickness of a stratigraphic unit

isoseismal line   line on the earth's surface connecting points of equal earthquake intensity

isotherm   line connecting points of equal temperature

jetty   artificial structure jutting outward from shore and designed to modify incoming currents

joint   one of a series of parallel breaks perpendicular to layering in rocks and along which there has been no movement

lahar   volcanic mudflow

lamination   thin layering within a bed of sedimentary rock

landslide   mass movement; movement of soil and rock down a slope

latitude   distance north or south from the equator, from zero to 90°; one degree is about 66 miles

lava   magma that flows out onto the earth's surface

levee   natural or artificial embankment on the borders of a stream that protects surrounding land from floods

limestone   sedimentary rock composed of calcium carbonate

liquefaction   transformation of soil or sediment from a solid to a liquid condition because of increased pore pressure

lithic sandstone   sandstone having less than 90% quartz and more rock fragments than feldspar

lithology   the characteristics of a rock, such as mineral composition and texture

longitude   distance east or west from the Prime Meridian, from 0 to 180° in either direction; 1° is about 66 miles

luster   the appearance of reflected light from a mineral surface

L-wave   earthquake wave that travels along surfaces rather than through the body of the earth

magma   molten silicate material from beneath the earth's surface

magnetic declination   angular difference between the geographic pole and the magnetic pole at the earth's surface

magnetic pole   exit point on the earth's surface of imaginary line through the earth's center that marks the axis of its magnetic field

magnitude   measure of the energy released by an earthquake

map scale   ratio between linear distance on a map and the true distance on the ground

mass wasting   synonym of landsliding

meander   freely developed bend or loop in a stream course

Mercalli scale   scale of earthquake intensity based on damage to manmade structures

metamorphic   rock formed by recrystallization of minerals without melting at high temperature and pressure

mineral   naturally occurring crystalline solid with a fixed chemical composition

montmorillonite   type of clay mineral that expands when wet

mudrock   sedimentary rock composed of grains less than 62 micrometers in size

nonhydrostatic   pressure not equal in all directions

nuee ardente   gaseous cloud of debris erupted from a volcano and moving rapidly downslope

ore mineral   mineral from which a valuable metal can be extracted at a profit

original horizontality   principle that sedimentary rocks are deposited with their surfaces parallel to the ground

outcrop   exposure of rock at the earth's surface

ozone   molecule composed of three atoms of oxygen

pedalfer   soil type in which the B-horizon is enriched in hematite

pedocal   soil type in which the B-horizon is enriched in calcium carbonate

perennial stream   stream that flows continuously throughout the year

permeability   ease of flow of fluid through a rock

phosphorite   sedimentary rock composed of chemically precipitated apatite

photosynthesis   chemical process in plants by which carbon dioxide and water are stimulated by sunlight to produce carbohydrates for plant nourishment

pH scale   logarithmic scale used to determine acidity or basicity of a liquid

phyllite   low-grade metamorphic rock composed largely of well-oriented sheet-structure minerals coarse enough to produce a shiny appearance in reflected light

placer deposit surficial mineral deposit formed by mechanical concentration of heavy minerals by wind or water

plankton microscopic one-celled organisms that live at the ocean surface

pollution level of contamination so high it has a harmful effect

pore cavity in a rock

porosity volume of space in a rock not occupied by grains

principal meridian a north-south line on the earth's surface around which a rectangular grid is established

Public Land Survey System rectangular grid system that divides most states into squares for surveying purposes

P-wave compressional earthquake wave that travels through the body of the earth

pyroclastic fragmental material hurled from a volcano during an eruption

quartz sandstone sandstone composed of more than 90% quartz

quick clay clay that loses its strength and behaves like a liquid after being disturbed

quicksand sand that loses its strength and behaves like a liquid after being disturbed

radioactive atom that undergoes spontaneous decay of its nucleus

radon colorless, odorless, tasteless radioactive gas produced by decay of uranium atoms

range series of six mile by six miles squares aligned north-south in the Public Land Survey System

Rayleigh wave earthquake wave that travels along surfaces rather than through the body of the earth

relief unevenness of the earth's surface, including differences in elevation and slope

reserves amount of valuable material whose location is known and that can be recovered at a profit

resources reserves plus other deposits that may someday be recoverable at a profit

salinization accumulation of salt in a soil that makes it unproductive

saltation hopping motion of sediment as it moves downcurrent in a stream

sandstone fragmental rock composed of grains between 2 mm and 0.06 mm in size

sanitary landfill site where solid waste is buried in a way that will not degrade the environment

schist metamorphic rock characterized by abundant oriented micas

seawall embankment along high-tide line on a shore that temporarily prevents wave erosion; usually artificial

secondary recovery pumping water into an oil well to recover additional petroleum

section one-square-mile subdivision of a township in the Public Land Survey System

shale fissile mudrock

slag frothy glassy artificial rock formed by quenching of molten material coming from a smelter

slate low-grade metamorphic rock formed from shale and having planar surfaces

smog ozone at ground level formed by chemical reaction between nitrogen oxides, hydrocarbon fumes from motor vehicles, and sunlight

soil surficial sedimentary material formed by chemical alteration of rocks plus organic matter

solubility amount of a substance that can be dissolved in a liquid at specified physical conditions

solution load dissolved material carried in a stream

soot black fine-grained carbon formed by incomplete burning of carbonaceous material

sorting a measure of the variation in grain sizes in a fragmental sediment

specific gravity weight of a substance compared to the weight of an equal volume of water

spreading failure ground subsidence caused by liquefaction

spring natural flow of water onto the ground or into a body of water from within a permeable rock

streak color of a mineral powder

stream capacity amount of sediment that can be carried by a stream

stream gradient amount of decrease in elevation per unit of horizontal distance, as measured in the direction of flow

stream terrace flat surface above the level of a stream, marking the level of a former floodplain

strike compass direction of a line formed by the intersection of a dipping surface with a horizontal plane

strip mine surface mine

subsoil B-horizon of a soil

Superfund fund of money established by the federal government to clean up the nation's most polluted sites

superposition "law" stating that in an undisturbed sequence of layered rocks the oldest ones are at the bottom

suspended load sediment carried within the body of stream water, normally silt and clay

S-waves surface waves generated by an earthquake

swelling soil soil that increases in volume when wet because it contains swelling clays

syncline concave upward fold in a rock layer

taconite banded iron ore as found in the Great Lakes region

tectonic the forces involved in deforming rocks

tertiary recovery pumping liquids or gases other than water down an oil well to increase the amount of oil produced

topographic map map that shows the shape of the land surface by means of contour lines

topographic profile the outline produced where the plane of a vertical slice intersects the ground surface

topsoil A-horizon

township series of six mile by six mile squares aligned east-west in the Public Land Survey System

toxic poisonous

tuff lithified volcanic ash

ultraviolet radiation wavelengths between 0.4 micrometers and 0.1 micrometers

unconformity buried surface of erosion

vertical exaggeration amount of increase in the vertical scale over the horizontal scale in a cross-section drawing

vertisol swelling soil

watershed area from which a stream gathers its water

water table top of the saturated zone in an unconfined aquifer

water witcher person who attempts to locate groundwater using a forked stick or piece of metal; also called dowser

wavelength distance between crests of a wave

C R E D I T S

Photos

Openers
© Doug Sherman/Geofile

Exercise 1
1.1: © Penny Tweedie/Tony Stone Images; **1.2:** © Jack Dermid/Photo Researchers, Inc.; **1.4:** Photo obtained from the Hong Kong Government, Geotechnical Control Office; **1.6:** © Christian Bossu-Pica/Tony Stone Images; **1.7:** Courtesy Barbara Smith, photo by Rachel Nalumoso

Exercise 2
Page 14 (plagioclase and orthoclase): © The McGraw-Hill Companies, Inc., Robert Rutford & James Zumberge/James Carter, photographer; **Page 14 (quartz and hornblende):** Harvey Blatt; **Page 14 (muscovite mica):** Photo by C.C. Plummer; **Page 15 (augite and garnet):** Harvey Blatt; **Page 15 (gypsum):** Photo by C.C. Plummer; **Page 15 (calcite):** © The McGraw-Hill Companies, Inc., Robert Rutford & James Zumberge/James Carter, photographer; **Page 15 (halite):** © The McGraw-Hill Companies, Inc./Bob Coyle, photographer; **Page 15 (pyrite):** © The McGraw-Hill Companies, Inc., Robert Rutford & James Zumberge/James Carter, photographer; **Page 16 (galena, fluorite):** Harvey Blatt

Exercise 3
Page 20 (top): © The McGraw-Hill Companies, Inc./Omni Resources; **Page 20 (bottom), Page 21 (all):** © The McGraw-Hill Companies, Inc., Robert Rutford and James Zumberge/James Carter, photographer; **3.5A:** Harvey Blatt; **3.5 B, C, D:** Photo by David McGeary; **3.6:** © D. Cavagnaro/Visuals Unlimited; **3.7A:** © The McGraw-Hill Companies, Inc., Robert Rutford and James Zumberge/James Carter, photographer; **3.7B:** Harvey Blatt; **3.8:** © St. Petersburg Times/Gamma-Liason; **3.9:** © The McGraw-Hill Companies, Inc./Omni Resources; **3.10A:** Photo Courtesy H.L. James; **3.10B:** © The McGraw-Hill Companies, Inc., Robert Rutford & James Zumberge/James Carter, photographer; **3.11 A, B:** Harvey Blatt; **3.11 C, D, E:** © The McGraw-Hill Companies, Inc., Robert Rutford and James Zumberge/James Carter, photographer; **3.11 F, G:** Photo by C.C. Plummer

Exercise 4
4.1: U.S. Geological Survey; **Page 38 (top):** U.S. Geological Survey

Exercise 5
5.7A: U.S. Geological Survey

Exercise 6
6.1: © Joyce Photographic/Photo Researchers, Inc.

Exercise 7
7.2: Harvey Blatt; **7.3:** Courtesy Ed Nuhfer, American Institute of Professional Geologists

Exercise 8
8.3: © Doug Sherman/Geofile; **8.51, 8.52:** U.S. Geological Survey

Exercise 9
9.7: U.S. Geological Survey

Exercise 10
10.6: U.S. Geological Survey

Exercise 11
11.3: N.H. Darton U.S. Geological Survey Water-Supply Paper 227; photo courtesy R.E. Fidler.; **11.4A:** U.S. Bureau of Reclamation; **11.4B:** From A.N. Palmer, *GSA Bulletin,* 103-1: Jan. 1991 (cover photo); **11.8:** Photo courtesy of Orlando Sentinel Star; **11.10B:** U.S. Geological Survey

Exercise 13
13.1: © Doug Sherman/Geofile

Exercise 14
14.1B: © James R. McCullagh/Visuals Unlimited; **14.7, 14.8:** U.S. Geological Survey

Exercise 15
15.1: © David J. Cross/Peter Arnold, Inc.

Exercise 16
16.6A: Courtesy of Eric Fantaine; **16.6B:** World Perspectives

Exercise 17
17.3: © H.W. Robinson/Visuals Unlimited; **17.4:** Indiana Geological Survey; **17.5:** U.S. Geological Survey; **17.6:** Photo courtesy of Kennecott Corporation © Don Green; **17.7:** © Don Duckson/Visuals Unlimited

Exercise 18
18.1: © Keith Wood/Tony Stone Images; **18.3:** The Department of Oil Properties, City of Long Beach.

Exercise 19
19.2: Photo by David McGeary; **19.5:** © David H. Ellis/Visuals Unlimited

Exercise 20
20.3 A,B: © Rafael Macia/Photo Researchers, Inc.

Exercise 21
21.2: © The McGraw-Hill Companies, Inc./Doug Sherman, photographer; **21.5:** © Ann Duncan/Tom Stack & Associates